LIAL VIDEO LIBRARY
WORKBOOK WITH
INTEGRATED REVIEW

CHRISTINE VERITY

INTERMEDIATE ALGEBRA

TWELFTH EDITION

Margaret L. Lial
American River College

John Hornsby
University of New Orleans

Terry McGinnis

PEARSON

Boston   Columbus   Indianapolis   New York   San Francisco
Amsterdam   Cape Town   Dubai   London   Madrid   Milan   Munich   Paris   Montreal   Toronto
Delhi   Mexico City   São Paulo   Sydney   Hong Kong   Seoul   Singapore   Taipei   Tokyo

ISBN-13: 978-0-13-428092-9
ISBN-10: 0-13-428092-X

1 2 3 4 5 6 RRD-H 19 18 17 16 15

www.pearsonhighered.com

CONTENTS for INTEGRATED REVIEW WORKSHEETS

CONTENTS for LIAL VIDEO LIBRARY WORKBOOK

Chapter 1 Linear Equations, Inequalities, and Applications

Learning Objectives
> Use the distributive property.
> Solve applied problems involving evaluating algebraic expressions.
> Use the rules for the order of operations.
> Know the meanings of $\neq$, $<$, $>$, $\leq$ and $\geq$.
> Solve applied problems involving inequalities.
> Classify numbers and graph them on number lines.
> Find the additive inverse of a real number.
> Find the absolute value of a real number.
> Interpret meaning of real numbers from a table of data.

Key Terms

Use the vocabulary terms listed below to complete each statement in exercises 1−14.

natural numbers	whole numbers	number line	additive inverse
integers	negative number	positive number	signed numbers
rational number	set-builder notation		coordinate
irrational number	real numbers	absolute value	

1. The set {0, 1, 2, 3, …} is called the set of _____.

2. The _____of a number is the same distance from 0 on the number line as the original number, but located on the opposite side of 0.

3. The whole numbers together with their opposites and 0 are called _____.

4. The set { 1, 2, 3, …} is called the set of _____.

5. The _____of a number is the distance between 0 and the number on the number line.

6. A _____ shows the ordering of the real numbers on a line.

7. A real number that is not a rational number is called a(n) _____.

8. The number that corresponds to a point on the number line is the _____ of that point.

9. A number located to the left of 0 on a number line is a _____.

10. A number located to the right of 0 on a number line is a _____.

11.   Numbers that can be represented by points on the number line are
        _____.

12.   _____ uses a variable and a description to describe a set.

13.   A number that can be written as the quotient of two integers is a
        _____.

14.   Positive numbers and negative numbers are _____.

Objective   Use the distributive property.

Video Examples

Review these examples:
Use the distributive property to rewrite each expression.

$7(p-6)$

$$7(p-6) = 7[p+(-6)]$$
$$= 7p + 7(-6)$$
$$= 7p - 42$$

$-3(5x-2)$

$$-3(5x-2) = -3[5x+(-2)]$$
$$= -3(5x)+(-3)(-2)$$
$$= (-3\cdot 5)x+(-3)(-2)$$
$$= -15x+6$$

$4\cdot 8 + 4\cdot 7$

$$4\cdot 8 + 4\cdot 7 = 4(8+7)$$

$5\cdot 3 + 5x + 5m$

$$5\cdot 3 + 5x + 5m = 5(3+x+m)$$

Now Try:
Use the distributive property to rewrite each expression.

$17(x-6)$

———————

$-4(2x-5)$

———————

$3\cdot 11 + 3\cdot 7$

———————

$12y + 12\cdot 6 + 12x$

———————

Rewrite each expression.

$-(5x+7)$

$$-(5x+7) = -1\cdot(5x+7)$$
$$= -1\cdot 5x + (-1)\cdot 7$$
$$= -5x - 7$$

$-(-p-5r+9x)$

$$-(-p-5r+9x)$$
$$= -1\cdot(-1p-5r+9x)$$
$$= -1\cdot(-1p)-1\cdot(-5r)-1\cdot(9x)$$
$$= p + 5r - 9x$$

Rewrite each expression.

$-(3x+4)$

———————

$-(-4x-5y+z)$

———————

1-3

$6a + 6b + 6$

$6a + 6b + 6 = 6a + 6b + 6 \cdot 1$

$\qquad = 6(a + b + 1)$

$3x + 3y + 3$

_____

Practice Exercises

Use the distributive property to rewrite each expression. Simplify if possible.

1. $n(2a - 4b + 6c)$

1. _____

2. $-2(5y - 9z)$

2. _____

3. $-(-2k + 7)$

3. _____

Objective   Solve applied problems involving evaluating algebraic expressions.

Practice Exercises

*A **mathematical model** is an equation that describes the relationship between two quantities. For example, the left expectancy at birth of Americans can be approximated by the equation*

$$y = 0.180x - 283.$$

where x is a year between 1960 and 2010 and y is age in years (Source: Centers for Disease Control and Prevention.)

　　　Use this model to approximate life expectancy (to the nearest year) in each of the following years.

4.　　1965　　　　　　　　　　　　　　　　　　**4.** _____

5.　　1990　　　　　　　　　　　　　　　　　　**5.** _____

6.　　2005　　　　　　　　　　　　　　　　　　**6.** _____

Objective   **Use the rules for the order of operations.**

Video Examples

Review this example:	**Now Try:**
Find the value of the expression.	Find the value of the expression.
$6+7\cdot3$	$5+2\cdot9$
Multiply, then add.	
$6+7\cdot3 = 6+21$	_____
$ = 27$	

Practice Exercises

Find the value of each expression.

7. $20 \div 5 - 3\cdot1$ **7.** _____

8. $3\cdot5^2 - 3\cdot7 - 9$ **8.** _____

9. $6^2 \div 3^2 - 4\cdot3 - 2\cdot5$ **9.** _____

Objective   **Know the meanings of ≠, <, >, ≤, and ≥.**

Video Examples

Review this example:
Determine whether the statement is true or false.

$$5 \cdot 6 - 12 \le 22$$

$$5 \cdot 6 - 12 \le 22$$

$$30 - 12 \le 22$$

$$18 \le 22$$

The statement is true.

Now Try:
Determine whether the statement is true or false.

$$7 \cdot 4 - 15 \le 13$$

_____

Practice Exercises

Tell whether each statement is true *or* false.

10. $3 \cdot 4 \div 2^2 \ne 3$ **10.** _____

11. $3.25 > 3.52$ **11.** _____

12. $2\left[7(4) - 3(5)\right] \le 45$ **12.** _____

Objective   **Solve applied problems involving inequalities.**

Practice Exercises

The table shows the number of pupils per teacher in U.S. public schools in selected states.

State	Pupils per Teacher
Alaska	16.2
Texas	14.7
California	24.1
Wyoming	12.5
Maine	12.3
Idaho	17.6
Missouri	13.8

Source: National Center for Education Statistics.

13.   Which states had a number greater than 14.7?   **13.** _____

14.   Which states had a number that was at most 16.2?   **14.** _____

15.   Which states had a number not less than 17.6?   **15.** _____

Objective   Classify numbers and graph them on number lines.

Video Examples

Review these examples:	**Now Try:**
Use an integer to express the boldface italic number in the application.	Use an integer to express the boldface italic number in the application.
In August, 2012, the National Debt was approximately $*16* trillion.	Death Valley is *282* feet below sea level.
Use –$16 trillion because "debt" indicates a negative number.	_____

Graph each number on a number line.

$$-3\frac{1}{2},\ -\frac{3}{2},\ 0,\ \frac{7}{2},\ 1$$

To locate the improper fractions on the number line, write them as mixed numbers or decimals.

Graph each number on a number line.

$$\frac{1}{2},\ 0,\ -3,\ -\frac{5}{2}$$

List the numbers in the following set that belong to each set of numbers.

$$\left\{-6,\ -\frac{5}{6},\ 0,\ 0.\overline{3},\ \sqrt{3},\ 4\frac{1}{5},\ 6,\ 6.7\right\}$$

Whole numbers

Answer: 0 and 6

Integers

Answer: –6, 0, and 6

Rational numbers

Answer: $-6,\ -\frac{5}{6},\ 0,\ 0.\overline{3},\ 4\frac{1}{5},\ 6,\ 6.7$

Irrational numbers

Answer: $\sqrt{3}$

List the numbers in the following set that belong to each set of numbers.

$$\left\{-10, -\frac{5}{8},\ 0,\ 0.\overline{4},\ \sqrt{5},\ 5\frac{1}{2},\ 7,\ 9.9\right\}$$

Whole numbers

_____

Integers

_____

Rational numbers

_____

Irrational numbers

_____

1-9

Practice Exercises

Use a real number to express each number in the following applications.

16. Last year Nina lost 75 pounds.

16. _____

17. Between 1970 and 1982, the population of Norway increased by 279,867.

17. _____

Graph the group of rational numbers on a number line.

18. −4.5, −2.3, 1.7, 4.2

18.

Objective   Find the additive inverse of a real number.

Practice Exercises

Find the additive inverse of each number.

19. -25 19. _____

20. $\frac{3}{8}$ 20. _____

21. 4.5 21. _____

Objective   **Find the absolute value of a real number.**

Video Examples

Review these examples:
Simplify by finding the absolute value.

$$|16|$$

$$|16| = 16$$

$$|-16|$$

$$|-16| = -(-16) = 16$$

$$-|-16|$$

$$-|-16| = -(16) = -16$$

Now Try:
Simplify by finding the absolute value.

$$|-10|$$

_____

$$-|10|$$

_____

$$|10-7|$$

_____

Practice Exercises

Simplify.

22. $-|49 - 39|$

23. $|-7.52 + 6.3|$

24. $|16 - 14|$

22. _____

23. _____

24. _____

Objective   **Interpret meaning of real numbers from a table of data.**

Video Examples

Review this example:

In the table, which category represents a decrease for both years?

Category	Change from 2012 to 2013	Change from 2013 to 2014
Eggs	−0.3	3.9
Milk	1.6	0.7
Orange Juice	−8.8	−3.9
Electricity	0.8	3.9

Source: U.S. Bureau of Labor and Statistics

Since a decrease implies a negative number, the category Orange Juice has a negative number for both years. So the answer is Orange Juice.

Now Try:

In the table to the left, which category represents an increase for both years?

_____

Practice Exercises

The Consumer Price Index (CPI) measures the average change in prices of goods and services purchased by urban consumers in the United States. The table shows the percent change in CPI for selected categories of goods and services from 2012 to 2013 and from 2013 to 2014. Use the table to answer each question.

Category	Change from 2012 to 2013	Change from 2013 to 2014
Gasoline	−1.2	−0.9
Eggs	−0.3	3.9
Milk	1.6	0.7
Electricity	0.8	3.9

25.   Which category represents a decrease for both years?   25. _____

26.   Which category in which year represents the greatest percent decrease?   26. _____

27.   Which category in which year represents the least change?   27. _____

Chapter 2 Linear Equations, Graphs, and Functions

Learning Objectives

Decide whether a given ordered pair is a solution of a given equation.
Complete ordered pairs for a given equation.

Key Terms

Use the vocabulary terms listed below to complete each statement in exercises 1–3.

ordered pair   **linear equation in two variables**   **table of values**

1. A table showing selected ordered pairs of numbers that satisfy an equation is called a _____.

2. A pair of numbers written between parentheses in which order is important is called a(n) _____.

3. An equation that can be written in the form $Ax + By = C$, where A, B, and C are real numbers and $A, B \neq 0$, is called a _____.

Objective   **Decide whether a given ordered pair is a solution of a given equation.**

Video Examples

Review these examples:

Decide whether each ordered pair is a solution of the equation $4x + 5y = 40$.

$(5,\ 4)$

Substitute 5 for x and 4 for y in the given equation.
$$4x + 5y = 40$$
$$4(5) + 5(4) \overset{?}{=} 40$$
$$20 + 20 \overset{?}{=} 40$$
$$40 = 40 \quad \text{True}$$
This result is true, so $(5,\ 4)$ is a solution of $4x + 5y = 40$.

$(-3,\ 6)$

Substitute –3 for x and 6 for y in the given equation.
$$4x + 5y = 40$$
$$4(-3) + 5(6) \overset{?}{=} 40$$
$$-12 + 30 \overset{?}{=} 40$$
$$18 = 40 \quad \text{False}$$
This result is false, so $(-3,\ 6)$ is not a solution of $4x + 5y = 40$.

Now Try:

Decide whether each ordered pair is a solution of the equation $4x + 5y = 40$.

$(8,\ 3)$

_____

$(5, -4)$

_____

Practice Exercises

Decide whether the given ordered pair is a solution of the given equation.

1. $4x - 3y = 10;\ (1, 2)$

1. _____

2. $2x - 3y = 1;\ \left(0, \frac{1}{3}\right)$

2. _____

3. $x = -7;\ (-7, 9)$

3. _____

Objective   **Complete ordered pairs for a given equation.**

Video Examples

Review this example:

Complete the ordered pair for the equation
$$y = 5x + 8.$$

$(3, \underline{\hspace{1cm}})$

Replace x with 3.
$$y = 5x + 8$$
$$y = 5(3) + 8$$
$$y = 15 + 8$$
$$y = 23$$
The ordered pair is $(3, 23)$.

Now Try:

Complete the ordered pair for the equation
$$y = 4x - 7.$$

$(5, \underline{\hspace{1cm}})$

_____

Practice Exercises

For each of the given equations, complete the ordered pairs beneath it.

4. $y = 2x - 5$

    (a) $(2,)$

    (b) $(0,)$

    (c) $(, 3)$

    (d) $(, -7)$

    (e) $(, 9)$

4.

    (a)   _____

    (b)   _____

    (c)   _____

    (d)   _____

    (e)   _____

5. $y = 3 + 2x$

    (a) $(-4,)$

    (b) $(2,)$

    (c) $(, 0)$

    (d) $(-2,)$

    (e) $(, -7)$

5.

    (a)   _____

    (b)   _____

    (c)   _____

    (d)   _____

    (e)   _____

Chapter 3 Systems of Linear Equations

Learning Objectives
   Evaluate algebraic expressions, given values for the variables.

Key Terms

Use the vocabulary terms listed below to complete each statement in exercises 1−3.

  variable   **constant**   **algebraic expression**

1. A _____ is a symbol, usually a letter, used to represent
   an unknown number.

2. A collection of numbers, variables, operation symbols, and grouping symbols is
   an_____.

3. A _____ is a fixed, unchanging number.

Objective   **Evaluate algebraic expressions, given values for the variables.**

Video Examples

Review these examples:
Find the value of each algebraic expression for
$x = 4$ and then $x = 7$.

$$5x^2$$

For $x = 4$,
$$5x^2 = 5 \cdot 4^2 \quad \text{Let } x = 4.$$
$$= 5 \cdot 16 \quad \text{Square 4.}$$
$$= 80 \quad \text{Multiply.}$$

For $x = 7$,
$$5x^2 = 5 \cdot 7^2 \quad \text{Let } x = 7.$$
$$= 5 \cdot 49 \quad \text{Square 7.}$$
$$= 245 \quad \text{Multiply.}$$

Now Try:
Find the value of each algebraic
expression for $x = 6$ and then
$x = 9$.
$$7x^2$$

_____

_____

Find the value of the expression for
$x = 7$ and $y = 6$.

$$3x + 4y + 2$$

   Replace x with 7 and y with 6.
$$3x + 4y + 2 = 3 \cdot 7 + 4 \cdot 6 + 2$$
$$= 21 + 24 + 2 \quad \text{Multiply.}$$
$$= 47 \quad \text{Add.}$$

Find the value of the expression for
$x = 8$ and $y = 4$.

$$5x + 6y + 1$$

_____

Practice Exercises

Find the value of each expression if $x = 2$ and $y = 4$.

1. $9x - 3y + 2$

1. _____

2. $\dfrac{2x + 3y}{3x - y + 2}$

2. _____

3. $\dfrac{3y^2 + 2x^2}{5x + y^2}$

3. _____

Chapter 4 Exponents, Polynomials, and Polynomial Functions

Learning Objectives
Use exponents.

Key Terms

Use the vocabulary terms listed below to complete each statement in exercises 1–3.

exponential expression   **base**   **power**

1. 2^5 is read "2 to the fifth _____".

2. A number written with an exponent is called a(n) _____.

3. The _____ is the number being multiplied repeatedly.

Objective   **Use exponents.**

Video Examples

Review these examples:

Write $5 \cdot 5 \cdot 5$ in exponential form.

Since 5 occurs as a factor three times, the base is 5 and the exponent is 3.

$5 \cdot 5 \cdot 5 = 5^3$

Now Try:

Write $4 \cdot 4 \cdot 4 \cdot 4 \cdot 4$ in exponential form.

_____

Name the base and exponent of each expression. Then evaluate.

3^4

Base: 3
Exponent: 4
Value: $3^4 = 3 \cdot 3 \cdot 3 \cdot 3 = 81$

$(-3)^4$

Base: –3
Exponent: 4
Value: $(-3)^4 = (-3)(-3)(-3)(-3) = 81$

Name the base and exponent of each expression. Then evaluate.

2^6

_____

$(-2)^6$

_____

Practice Exercises

Write the expression in exponential form and evaluate, if possible.

1. $\left(\frac{1}{3}\right)\left(\frac{1}{3}\right)\left(\frac{1}{3}\right)\left(\frac{1}{3}\right)\left(\frac{1}{3}\right)$

1. _____

Evaluate each exponential expression. Name the base and the exponent.

2. $(-4)^4$

2. _____

base _____

exponent _____

3. -3^8

3. _____

base _____

exponent _____

Chapter 5 Factoring

Learning Objectives
Multiply any two polynomials.
Multiply binomials.

Key Terms

Use the vocabulary terms listed below to complete each statement in exercises 1–3.

> **FOIL**   **outer product**   **inner product**

1. The _____ of $(2y-5)(y+8)$ is $-5y$.

2. _____ is a shortcut method for finding the product of two binomials.

3. The _____ of $(2y-5)(y+8)$ is $16y$.

Objective   **Multiply any two polynomials.**

Video Examples

Review these examples:
  Find each product.

$$6x^3(-5x^3+7x-8)$$

$$6x^3(-5x^3+7x-8)$$
$$= 6x^3(-5x^3)+6x^3(7x)+6x^3(-8)$$
$$= -30x^6+42x^4-48x^3$$

$$5x^3(x+4)(x-7)$$
$$5x^3(x+4)(x-7)$$
$$= 5x^3\left[(x+4)(x)+(x+4)(-7)\right]$$
$$= 5x^3\left[x^2+4x-7x-28\right]$$
$$= 5x^3(x^2-3x-28)$$
$$= 5x^5-15x^4-140x^3$$

Now Try:
  Find each product.

$$7x^2(-4x^2-8x-6)$$

_____

$$3x^2(x-1)(x+6)$$

_____

Practice Exercises

Find each product.

1. $7b^2(-5b^2+1-4b)$

2. $(3m-5)(2m+4)$

3. $3m^3+2m^2-4m$
   $$\underline{\hspace{2em} 2m^2+1}$$

1. _____

2. _____

3. _____

Objective   Multiply binomials.

Video Examples

Review this example:
Use the FOIL method to find the product.

$$(9x - 4y)(7x + 6y)$$

$$(9x - 4y)(7x + 6y)$$

    First   Outer   Inner   Last

$$= 63x^2 + 54xy - 28xy - 24y^2$$

$$= 63x^2 + 26xy - 24y^2$$

Now Try:
Use the FOIL method to find the product.

$$(6z - 4)(9z - 5)$$

_____

Practice Exercises

Find each product.

4. $(3x + 2y)(2x - 3y)$ 4. _____

5. $(5a - b)(4a + 3b)$ 5. _____

6. $(x - 5)(x + 3)$ 6. _____

Chapter 6 Rational Expressions and Functions

Learning Objectives
Write a fraction in lowest terms using a common factor.
Multiply signed fractions.
Divide signed fractions.
Add and subtract unlike fractions.
Rewrite mixed numbers as improper fractions, or the reverse.
Solve application problems containing mixed numbers.

Key Terms

Use the vocabulary terms listed below to complete each statement in exercises 1−8.

> **equivalent fractions**   **common factor**   **lowest terms**
>
> **like fractions**   **unlike fractions**   **least common denominator**
>
> **mixed number**   **improper fraction**

1. A fraction is written in _____ when its numerator and denominator have no common factor other than 1.

2. A _____ is a number that can be divided into two or more whole numbers.

3. Two fractions are _____ when they represent the same portion of a whole.

4. A(n) _____ includes a fraction and a whole number written together.

5. A mixed number can be rewritten as a(n) _____.

6. Fractions with different denominators are called _____.

7. Fractions with the same denominator are called _____.

8. The _____ of two whole numbers is the smallest whole number divisible by both of the numbers.

Objective   Write a fraction in lowest terms using a common factor.

Video Examples

Review these examples:
  Write each fraction in lowest terms.

$$\frac{32}{48}$$

Divide both numerator and denominator by 16.
$$\frac{32}{48} = \frac{32 \div 16}{48 \div 16} = \frac{2}{3}$$

$$\frac{28}{56}$$

Suppose we thought that 4 was the greatest common factor of 28 and 56. Dividing by 4 would give
$$\frac{28}{56} = \frac{28 \div 4}{56 \div 4} = \frac{7}{14}$$

But $\frac{7}{14}$ is not in lowest terms, because 7 and 14 have a common factor of 7. So we divide by 7.
$$\frac{7}{14} = \frac{7 \div 7}{14 \div 7} = \frac{1}{2}$$

The fraction $\frac{28}{56}$ could have been written in lowest terms in one step by dividing by 28, the greatest common factor of 28 and 56.
$$\frac{28}{56} = \frac{28 \div 28}{56 \div 28} = \frac{1}{2}$$

Now Try:
  Write each fraction in lowest terms.

$$\frac{18}{27}$$

_____

$$\frac{27}{45}$$

_____

Practice Exercises

Write each fraction in lowest terms.

1. $\dfrac{14}{49}$

2. $\dfrac{8}{36}$

3. $\dfrac{30}{42}$

1. _____

2. _____

3. _____

Objective   Multiply signed fractions.

Video Examples

Review these examples:

Multiply. Write the product in lowest terms.

$$-\frac{7}{9} \cdot -\frac{5}{11}$$

Multiply the numerators and multiply the denominators.

$$-\frac{7}{9} \cdot -\frac{5}{11} = \frac{7 \cdot 5}{9 \cdot 11} = \frac{35}{99}$$

The answer is in lowest terms because 35 and 99 have no common factor other than 1.

$$-\frac{9}{7}\left(\frac{7}{15}\right)$$

Multiplying a negative number times a positive number gives a negative product.

$$-\frac{9}{7}\left(\frac{7}{15}\right) = -\frac{3 \cdot 3 \cdot 7}{7 \cdot 3 \cdot 5} = -\frac{\overset{1}{\cancel{3}} \cdot 3 \cdot \overset{1}{\cancel{7}}}{\underset{1}{\cancel{7}} \cdot \underset{1}{\cancel{3}} \cdot 5} = -\frac{3}{5}$$

Find $\frac{3}{8}$ of $\frac{4}{9}$.

Recall that "of" indicates multiplication.

$$\frac{3}{8} \cdot \frac{4}{9} = \frac{3 \cdot 2 \cdot 2}{2 \cdot 2 \cdot 2 \cdot 3 \cdot 3} = \frac{\overset{1}{\cancel{3}} \cdot \overset{1}{\cancel{2}} \cdot \overset{1}{\cancel{2}}}{\underset{1}{\cancel{2}} \cdot \underset{1}{\cancel{2}} \cdot 2 \cdot \underset{1}{\cancel{3}} \cdot 3} = \frac{1}{6}$$

Now Try:

Multiply. Write the product in lowest terms.

$$-\frac{10}{11} \cdot -\frac{4}{13}$$

_____

$$-\frac{11}{7}\left(\frac{14}{33}\right)$$

_____

Find $\frac{2}{7}$ of $\frac{21}{40}$.

_____

Practice Exercises

Multiply. Write the products in lowest terms.

4. $-\frac{10}{42} \cdot \frac{3}{5}$

4. _____

5. $\frac{6}{18} \cdot \frac{9}{2}$

5. _____

6. $\frac{5}{9}$ of 81

6. _____

Objective   Divide signed fractions.

Video Examples

Review these examples:

Divide. Write each quotient in lowest terms.

$$\frac{5}{7} \div \frac{10}{3}$$

$$\frac{5}{7} \div \frac{10}{3} = \frac{5}{7} \cdot \frac{3}{10} = \frac{\overset{1}{\cancel{5}} \cdot 3}{7 \cdot 2 \cdot \underset{1}{\cancel{5}}} = \frac{3}{14}$$

$$6 \div \left(-\frac{1}{5}\right)$$

$$6 \div \left(-\frac{1}{5}\right) = \frac{6}{1} \cdot \left(-\frac{5}{1}\right) = -\frac{6 \cdot 5}{1 \cdot 1} = -\frac{30}{1} = -30$$

$$-\frac{7}{9} \div (-6)$$

$$-\frac{7}{9} \div (-6) = -\frac{7}{9} \cdot \left(-\frac{1}{6}\right) = \frac{7 \cdot 1}{3 \cdot 3 \cdot 3 \cdot 2} = \frac{7}{54}$$

Now Try:

Divide. Write each quotient in lowest terms.

$$\frac{5}{9} \div \frac{20}{3}$$

_____

$$10 \div \left(-\frac{1}{8}\right)$$

_____

$$-\frac{6}{13} \div \left(-\frac{3}{26}\right)$$

_____

Practice Exercises

Divide. Write the quotients in lowest terms.

7. $\frac{7}{8} \div (-21)$

7. _____

8. $-\frac{5}{12} \div \frac{15}{8}$

8. _____

9. $-\frac{2}{3} \div \left(-\frac{7}{9}\right)$

9. _____

Objective   Add and subtract unlike fractions.

Video Examples

Review these examples:	**Now Try:**

Review these examples:

Find each sum or difference.

$$\frac{1}{2}+\frac{1}{6}$$

Step 1 The larger denominator (6) is the LCD.

Step 2 $\frac{1}{2}=\frac{1\cdot 3}{2\cdot 3}=\frac{3}{6}$ and $\frac{1}{6}$ already has the LCD.

Step 3 Add the numerators.

$$\frac{1}{2}+\frac{1}{6}=\frac{3}{6}+\frac{1}{6}=\frac{3+1}{6}=\frac{4}{6}$$

Step 4 Write $\frac{4}{6}$ in lowest terms.

$$\frac{4}{6}=\frac{2\cdot \overset{1}{\cancel{2}}}{\cancel{2}\cdot 3}=\frac{2}{3}$$

$$\frac{3}{8}-\frac{7}{12}$$

Step 1 The LCD is 24.

Step 2 $\frac{3}{8}=\frac{3\cdot 3}{8\cdot 3}=\frac{9}{24}$ and $\frac{7}{12}=\frac{7\cdot 2}{12\cdot 2}=\frac{14}{24}$

Step 3 Subtract the numerators.

$$\frac{3}{8}-\frac{7}{12}=\frac{9}{24}-\frac{14}{24}=\frac{9-14}{24}=\frac{-5}{24}\text{, or }-\frac{5}{24}$$

Step 4 $-\frac{5}{24}$ is in lowest terms.

$$-\frac{7}{18}+\frac{5}{12}$$

Step 1 Use prime factorization to find the LCD.

$$18=2\cdot 3\cdot 3$$
$$12=2\cdot 2\cdot 3$$
$$\text{LCD}=2\cdot 2\cdot 3\cdot 3=36$$

Step 2

$$-\frac{7}{18}=-\frac{7\cdot 2}{18\cdot 2}=-\frac{14}{36}\text{ and }\frac{5}{12}=\frac{5\cdot 3}{12\cdot 3}=\frac{15}{36}$$

Step 3 Add the numerators.

$$-\frac{7}{18}+\frac{5}{12}=-\frac{14}{36}+\frac{15}{36}=\frac{-14+15}{36}=\frac{1}{36}$$

Step 4 $\frac{1}{36}$ is in lowest terms.

Now Try:

Find each sum or difference.

$$\frac{5}{9}+\frac{7}{18}$$

———————

$$\frac{8}{15}-\frac{7}{10}$$

———————

$$-\frac{5}{24}+\frac{7}{9}$$

———————

Practice Exercises

Find each sum or difference. Write all answers in lowest terms.

10. $\dfrac{1}{6} + \dfrac{2}{15}$

10. _____

11. $-\dfrac{1}{2} + \dfrac{7}{12}$

11. _____

12. $\dfrac{33}{40} - \dfrac{7}{24}$

12. _____

Objective   Rewrite mixed numbers as improper fractions, or the reverse.

Video Examples

Review these examples:

Write $9\frac{3}{4}$ as an improper fraction.

Step 1 $9\frac{3}{4}$ $4 \cdot 9 = 36$ Then $36 + 3 = 39$

Step 2 $9\frac{3}{4} = \frac{39}{4}$

Write the improper fraction as an equivalent mixed number in simplest form.

$\frac{18}{7}$

Divide 18 by 7.

$$7\overline{)18} \quad \begin{array}{r} 2 \\ \underline{14} \\ 4 \end{array}$$

The quotient 2 is the whole number part. The remainder 4 is the numerator of the fraction, and the denominator stays as 7.

$\frac{18}{7} = 2\frac{4}{7}$

Now Try:

Write $5\frac{5}{6}$ as an improper fraction.

_____

Write the improper fraction as an equivalent mixed number in simplest form.

$\frac{27}{8}$

_____

Practice Exercises

Write each mixed number as an improper fraction.

13. $-8\frac{2}{7}$

14. $-1\frac{7}{9}$

Write the improper fraction as a mixed number in simplest form.

15. $\frac{26}{3}$

13. _____

14. _____

15. _____

Objective   Solve application problems containing mixed numbers.

Video Examples

Review this example:

First, estimate the answer to the application problem. Then find the exact answer.

George's daughter grew $1\frac{1}{3}$ inches last year and $2\frac{1}{5}$ inches this year. How much has her height increased over the two years?

First, round each mixed number to the nearest whole number.

$1\frac{1}{3}$ rounds to 1   and $2\frac{1}{5}$ rounds to 2

Using the rounded numbers, we add.

$1 + 2 = 3 \leftarrow$ Estimate

To find the exact answer, use the original mixed numbers and add.

$$1\frac{1}{3} + 2\frac{1}{5} = \frac{4}{3} + \frac{11}{5} = \frac{20}{15} + \frac{33}{15} = \frac{20+33}{15}$$

$$= \frac{53}{15} = 3\frac{8}{15}$$

Her height increased $3\frac{8}{15}$ in. over the two years.

This result is close to the estimate of 3 in.

Now Try:

First, estimate the answer to the application problem. Then find the exact answer.

A plumber has three pieces of pipe measuring $2\frac{1}{5}$ ft, $3\frac{3}{4}$ ft, and $4\frac{1}{8}$ ft. What is the total length of pipe?

Estimate _____

Exact _____

Practice Exercises

First, estimate the answer to each application problem. Then find the exact answer.

16. A living room has dimensions $3\frac{3}{4}$ meters by $3\frac{1}{3}$ meters. What is the area of the room?

16.
Estimate_____

Exact _____

17. Suppose that a pair of pants requires $3\frac{1}{8}$ yd of material. How much material would be needed for 6 pairs of pants?

17.
Estimate_____

Exact _____

Chapter 7 Roots, Radicals, and Root Functions

Learning Objectives
  Find square roots.

Key Terms

Use the vocabulary terms listed below to complete each statement in exercises 1−5.

  square root   **principal square root**   **radicand**

  radical   **radical expression**

1.   The number or expression inside a radical sign is called the _____.

2.   The number b is a _____ of a if $b^2 = a$.

3.   The expression $\sqrt[n]{a}$ is called a _____.

4.   The positive square root of a number is its _____.

5.   A _____ is a radical sign and the number or expression in it.

Objective   Find square roots.

Video Examples

Review these examples:	Now Try:
Find the square roots of 64.	Find the square roots of 81.
What number multiplied by itself equals 64?	
$8^2 = 64$ and $(-8)^2 = 64$.	
Thus, 64 has two square roots: 8 and −8.	_____

Find each square root.	Find each square root.
$\sqrt{121}$	$\sqrt{169}$
$11^2 = 121$, so $\sqrt{121} = 11$.	
	_____
$-\sqrt{\dfrac{16}{25}}$	$-\sqrt{\dfrac{9}{49}}$
$-\sqrt{\dfrac{16}{25}} = -\dfrac{4}{5}$	
	_____

Find the square of each radical expression.

$$\sqrt{17}$$

The square of $\sqrt{17}$ is $\left(\sqrt{17}\right)^2 = 17$.

$$\sqrt{w^2 + 3}$$

$$\left(\sqrt{w^2 + 3}\right)^2 = w^2 + 3$$

Find the square of each radical expression.

$$\sqrt{19}$$

_____

$$\sqrt{n^2 + 5}$$

_____

Practice Exercises

Find all square roots of each number.

1. 625

2. $\dfrac{121}{196}$

Find the square root.

3. $\sqrt{\dfrac{900}{49}}$

1. _____

2. _____

3. _____

Chapter 8 Quadratic Equations, Inequalities, and Functions

Learning Objectives
      Simplify expressions.

Objective   Simplify expressions.

Video Examples

Review these examples:

Simplify each expression.

$8(4m - 6n)$

Use the distributive property.
$$8(4m - 6n) = 8(4m) + 8(-6n)$$
$$= 32m - 48n$$

$9 - (4y - 6)$

$$9 - (4y - 6) = 9 - 1(4y - 6)$$
$$= 9 - 4y + 6$$
$$= 15 - 4y$$

Now Try:

Simplify each expression.

$7(5x - 3y)$

_____

$8 - (7x - 3)$

_____

Practice Exercises

Simplify each expression.

1. $4(2x + 5) + 7$

1. _____

2. $-4 + s - (12 - 21)$

2. _____

3. $-2(-5x + 2) + 7$

3. _____

Chapter 9 Inverse, Exponential, and Logarithmic Functions

Learning Objectives
Understand the definition of a function.
Find domain and range.
Identify functions defined by graphs and equations.
Use function notation.

Key Terms

Use the vocabulary terms listed below to complete each statement in exercises 1−3.

function   **domain**   **range**

1. The _____ of a relation is the set of second components
    (y-values) of the ordered pairs of the relation.

2. A _____ is a set of ordered pairs in which each first
    component corresponds to exactly one second component.

3. The _____ of a relation is the set of first components
    (x-values) of the ordered pairs of the relation.

Objective   **Understand the definition of a function.**

Video Examples

Review these examples:
Determine whether each relation is a function.

  $\{(7, 4), (-7, 3), (7, 2)\}$

The first component 7 appears in two ordered
pairs and corresponds to two different second
components. Therefore, this relation is not a
function.

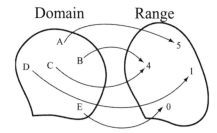

The points here include (A, 5), (B, 4), (C, 4),
(D, 1), and (E, 0).
Each first component appears once and only
once. The relation is a function.

Now Try:
Determine whether each relation is a
function.
  $\{(10, 0), (10, 5), (10, 20)\}$

_____

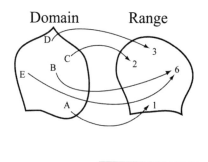

_____

Practice Exercises

Determine whether each relation is a function.

1. $\{(1,3),(5,7),(11,9),(8,-2),(6,-7),(-4,-3)\}$

   1. _____

2. $\{(-1.2,4),(1.8,-2.5),(3.7,-3.8),(3.7,3.8)\}$

   2. _____

3. $\{(-3,5),(-2,5),(-1,0),(0,-5),(1,5)\}$

   3. _____

Objective   Find domain and range.

Video Examples

Review these examples:
Give the domain and range of each relation. Tell whether the relation defines a function.

$\{(15, 2), (20, 3), (6, 10), (-1, 2)\}$

The domain is the set of x-values $\{-1, 6, 15, 20\}$. The range is the set of y-values $\{2, 3, 10\}$. The relation is a function, because each x-value corresponds to exactly one y-value.

Now Try:
Give the domain and range of each relation. Tell whether the relation defines a function.

$\{(13, -1), (13, -2), (13, 4)\}$

domain: _____

range: _____

_____

Give the domain and range of each relation.

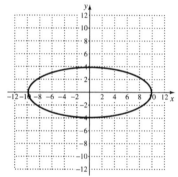

The x-values of the points on the graph include all numbers between -10 and 10, inclusive. The y-values include all numbers between -4 and 4, inclusive.
The domain is $[-10, 10]$. The range is $[-4, 4]$.

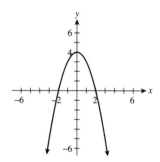

The graph extends indefinitely left and right, as well as downward. The domain is $(-\infty, \infty)$. Because there is a greatest y-value, 4, the range includes all numbers less than or equal to 4, written $(-\infty, 4]$.

Give the domain and range of each relation.

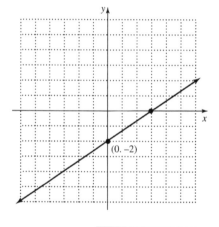

(0, -2)

_____

_____

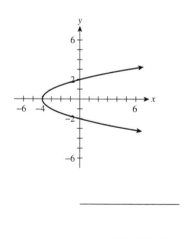

_____

_____

9-3

Practice Exercises

Decide whether the relation is a function, and give the domain and range of the relation.

4. $\{(5,\ 2),\ (3,-1),\ (1,-3),\ (-1,-5)\}$

4. _____

domain:_____

range: _____

5.

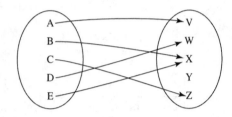

5. _____

domain:_____

range: _____

6.

x	y
1	3
2	−1
−1	4
1	4

6. _____

domain:_____

range: _____

Objective   Identify functions defined by graphs and equations.

Video Examples

Review these examples:

Use the vertical line test to determine whether the relation graphed is a function.

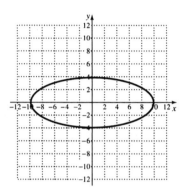

The graph is not a function.

Now Try:

Use the vertical line test to determine whether the relation graphed is a function.

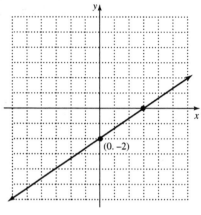

_____

Decide whether the relation defines y as a function of x. Give the domain.

$$y = 2x - 4$$

Each x value corresponds to just one y-value and the relation defines a function. Since x can be any real number, the domain is $(-\infty, \infty)$.

Decide whether the relation defines y as a function of x. Give the domain.

$$y = 3x - 1$$

_____

_____

Practice Exercises

Use the vertical line test to determine whether the relation graphed is a function.

7.

7. _____

Decide whether each equation defines y as a function of x. Give the domain.

8. $y^2 = x + 1$

8. _____

9. $y = \dfrac{3}{x+6}$

9. _____

Objective   Use function notation.

Video Examples

Review these examples:

Let $f(x) = 7x - 3$. Evaluate the function f for $x = 4$.

Start with the given function. Replace x with 4.
$$f(x) = 7x - 3$$
$$f(4) = 7(4) - 3$$
$$f(4) = 28 - 3$$
$$f(4) = 25$$
Thus, $f(4) = 25$.

Let $g(x) = 5x + 6$. Find and simplify $g(n + 8)$.

Replace x with $n + 8$.
$$g(x) = 5x + 6$$
$$g(n + 8) = 5(n + 8) + 6$$
$$g(n + 8) = 5n + 40 + 6$$
$$g(n + 8) = 5n + 46$$

For the function, find $f(5)$.
$$f = \{(7, -27),\ (5, -25),\ (3, -23),\ (1, -21)\}$$

From the ordered pair $(5, -25)$, we have $f(5) = -25$.

Write the equation using function notation $f(x)$. Then find $f(-5)$.
$$x - 5y = 8$$

Step 1 $x - 5y = 8$
$$-5y = -x + 8$$
$$y = \frac{1}{5}x - \frac{8}{5}$$

Step 2 $f(x) = \frac{1}{5}x - \frac{8}{5}$
$$f(-5) = \frac{1}{5}(-5) - \frac{8}{5}$$
$$f(-5) = -\frac{13}{5}$$

Now Try:

Let $f(x) = 8x - 7$. Evaluate the function f for $x = 3$.

_____

Let $g(x) = 4x - 7$. Find and simplify $g(a - 1)$.

_____

For the function, find $f(-6)$.
$$f = \{(-2, 11),\ (-4, 17),\ (-6, 21),\ (-8, 24)\}$$

_____

Write the equation using function notation $f(x)$. Then find $f(-3)$.
$$2x + 3y = 7$$

_____

9-7

Practice Exercises

For each function f, find (a) $f(-2)$, *(b)* $f(0)$, *and (c)* $f(-x)$.

10. $f(x) = 3x - 7$

10. a._____

b._____

c._____

11. $f(x) = 2x^2 + x - 5$

11. a._____

b._____

c._____

12. $f(x) = 9$

12. a._____

b._____

c._____

Chapter 10 Nonlinear Functions, Conic Sections, and Nonlinear Systems

Learning Objectives
Graph functions defined by radical expressions.
Solve linear systems by substitution.
Solve linear systems by elimination.
Graph linear inequalities in two variables.

Key Terms

Use the vocabulary terms listed below to complete each statement in exercises 1−8.

radical expression	**square root**	**cube root**
addition property of equality	**elimination method**	**substitution**
linear inequality in two variables	**boundary line**	

1. In the graph of a linear inequality, the _____ separates the region that satisfies the inequality from the region that does not satisfy the inequality.

2. The number b is a _____ of a if $b^3 = a$.

3. The expression $\sqrt[n]{a}$ is called a _____ .

4. The number b is a _____ of a if $b^2 = a$.

5. An inequality that can be written in the form $Ax + By < C$, $Ax + By > C$, $Ax + By \leq C$, or $Ax + By \geq C$ is called a_____.

6. Using the addition property to solve a system of equations is called the _____.

7. The _____ states that the same added quantity to each side of an equation results in equal sums.

8. _____ is being used when one expression is replaced by another.

Objective 1 Graph functions defined by radical expressions.

Video Examples

Review these examples:
Graph each function by creating a table of values. Give the domain and the range.

$$f(x) = \sqrt{x-1}$$

Create a table of values.

x	$f(x) = \sqrt{x-1}$
1	$\sqrt{1-1} = 0$
5	$\sqrt{5-1} = 2$
10	$\sqrt{10-1} = 3$

For the radicand to be nonnegative, we must have $x-1 \geq 0$ or $x \geq 1$. Therefore, the domain is $[1, \infty)$. Function values are nonnegative, so the range is $[0, \infty)$.

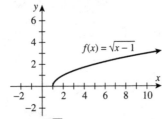

$$f(x) = \sqrt[3]{x} + 1$$

Create a table of values.

x	$f(x) = \sqrt[3]{x} + 1$
-8	$\sqrt[3]{-8} + 1 = -1$
-1	$\sqrt[3]{-1} + 1 = 0$
0	$\sqrt[3]{0} + 1 = 1$
1	$\sqrt[3]{1} + 1 = 2$
8	$\sqrt[3]{8} + 1 = 3$

Both the domain and range are $(-\infty, \infty)$.

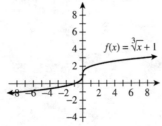

Now Try:
Graph each function by creating a table of values. Give the domain and the range.

$$f(x) = \sqrt{x} - 1$$

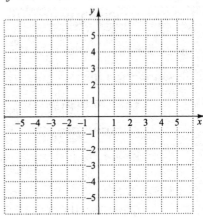

domain: _____

range: _____

$$f(x) = \sqrt[3]{x} + 1$$

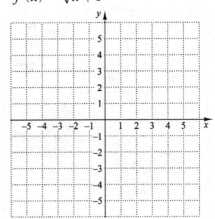

domain: _____

range: _____

Practice Exercises

Graph each function and give its domain and its range.

1. $f(x) = \sqrt{x} + 2$

1.

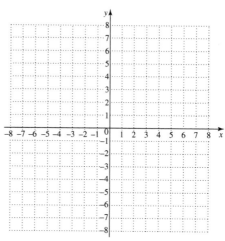

domain: _____

range: _____

2. $f(x) = \sqrt[3]{x} - 2$

2.

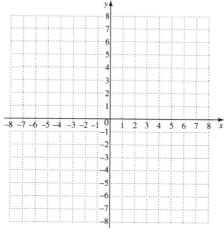

domain: _____

range: _____

3. $f(x) = \sqrt[3]{x} + 2$

3.

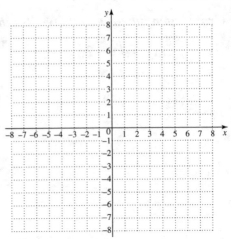

domain: _____

range: _____

Objective   **Solve linear systems by substitution.**

Video Examples

Review these examples:

Solve the system by the substitution method.
$$2x + 5y = 22 \quad (1)$$
$$y = 4x \quad (2)$$

Equation (2) is already solved for y. We substitute $4x$ for y in equation (1).
$$2x + 5y = 22$$
$$2x + 5(4x) = 22$$
$$2x + 20x = 22$$
$$22x = 22$$
$$x = 1$$

Find the value of y by substituting 1 for x in either equation. We use equation (2).
$$y = 4x$$
$$y = 4(1) = 4$$

We check the solution $(1, 4)$ by substituting 1 for x and 4 for y in both equations.

$$2x + 5y = 22 \qquad\qquad y = 4x$$
$$2(1) + 5(4) \overset{?}{=} 22 \qquad\qquad 4 \overset{?}{=} 4(1)$$
$$2 + 20 \overset{?}{=} 22 \qquad\qquad \text{True } 4 = 4$$
$$22 = 22 \text{ True}$$

Since $(1, 4)$ satisfies both equations, the solution set of the system is $\{(1, 4)\}$.

Solve the system by the substitution method.
$$4x + 5y = 13 \qquad (1)$$
$$x = -y + 2 \quad (2)$$

Equation (2) gives x in terms of y. We substitute $-y + 2$ for x in equation (1).
$$4x + 5y = 13$$
$$4(-y + 2) + 5y = 13$$
$$-4y + 8 + 5y = 13$$
$$y + 8 = 13$$
$$y = 5$$

Find the value of x by substituting 5 for y in either equation. We use equation (2).

Now Try:

Solve the system by the substitution method.
$$x + y = 7$$
$$y = 6x$$

_____

Solve the system by the substitution method.
$$2x + 3y = 6$$
$$x = 5 - y$$

_____

$$x = -y + 2$$

$$x = -5 + 2 = -3$$

We check the solution $(-3, 5)$ by substituting -3 for x and 5 for y in both equations.

$$4x + 5y = 13 \qquad\qquad x = -y + 2$$
$$4(-3) + 5(5) \overset{?}{=} 13 \qquad -3 \overset{?}{=} -5 + 2$$
$$-12 + 25 \overset{?}{=} 13 \qquad \text{True} \quad -3 = -3$$
$$13 = 13 \quad \text{True}$$

Both results are true, so the solution set of the system is $\{(-3, 5)\}$.

Solve the system by the substitution method.
$$4x = 5 - y \quad (1)$$
$$7x + 3y = 15 \qquad (2)$$

Step 1 Solve one of the equations for x or y. Solve equation (1) for y to avoid fractions.
$$4x = 5 - y$$
$$y + 4x = 5$$
$$y = -4x + 5$$

Step 2 Now substitute $-4x + 5$ for y in equation (2).
$$7x + 3y = 15$$
$$7x + 3(-4x + 5) = 15$$

Step 3 Solve the equation from Step 2.
$$7x - 12x + 15 = 15$$
$$-5x + 15 = 15$$
$$-5x = 0$$
$$x = 0$$

Step 4 Equation (1) solved for y is $y = -4x + 5$. Substitute 0 for x.
$$y = -4(0) + 5 = 5$$

Step 5 Check that $(0, 5)$ is the solution.
$$4x = 5 - y \qquad\quad 7x + 3y = 15$$
$$4(0) \overset{?}{=} 5 - 5 \qquad 7(0) + 3(5) \overset{?}{=} 15$$
$$0 = 0 \quad \text{True} \qquad \text{True} \quad 15 = 15$$

Since both results are true, the solution set of the system is $\{(0, 5)\}$.

Solve the system by the substitution method.
$$2x + 7y = 2$$
$$3y = 2 - x$$

_____

Practice Exercises

Solve each system by the substitution method. Check each solution.

4. $3x + 2y = 14$

       $y = x + 2$

4. _____

5. $x + y = 9$

     $5x - 2y = -4$

5. _____

6. $3x - 21 = y$

     $y + 2x = -1$

6. _____

Objective   **Solve linear systems by elimination.**

Video Examples

Review this example:

Use the elimination method to solve the system.

$$x + y = 6 \quad (1)$$
$$-x + y = 4 \quad (2)$$

Add the equations vertically.

$$x + y = 6 \quad (1)$$
$$-x + y = 4 \quad (2)$$
$$\overline{2y = 10}$$
$$y = 5$$

To find the x-value, substitute 5 for y in either of the two equations of the system. We choose equation (1).

$$x + y = 6$$
$$x + 5 = 6$$
$$x = 1$$

Check the solution (1, 5), by substituting 1 for x and 5 for y in both equations of the given system.

$x + y = 6$	$-x + y = 4$
$1 + 5 \overset{?}{=} 6$	$-1 + 5 \overset{?}{=} 4$
$6 = 6$ True	True $4 = 4$

Since both results are true, the solution set of the system is $\{(1, 5)\}$.

Now Try:

Use the elimination method to solve the system.

$$x + y = 11$$
$$x - y = 5$$

_____

Practice Exercises

Solve each system by the elimination method. Check your answers.

7. $x - 4y = -4$
       $-x + y = -5$

7. _____

8. $2x - y = 10$
$3x + y = 10$

8. _____

9. $x - 3y = 5$
$-x + 4y = -5$

9. _____

Objective   **Graph linear inequalities in two variables.**

Video Examples

Review these examples:

Graph $3x - 2y \leq 6$.

The inequality $3x - 2y \leq 6$ means that
$$3x - 2y < 6 \text{ or } 3x - 2y = 6.$$
We begin by graphing the line $3x - 2y = 6$ with intercepts $(0, -3)$ and $(2, 0)$. This boundary line divides the plane into two regions, one of which satisfies the inequality. We use the test point $(0, 0)$ to see whether the resulting statement is true or false, thereby determining whether the point is in the shaded region or not.

$$3x - 2y \leq 6$$
$$3(0) - 2(0) \overset{?}{\leq} 6$$
$$0 - 0 \overset{?}{\leq} 6$$
$$0 \leq 6 \quad \text{True}$$

Since the last statement is true, we shade the region that includes the test point $(0, 0)$. The shaded region, along with the boundary line, is the desired graph.

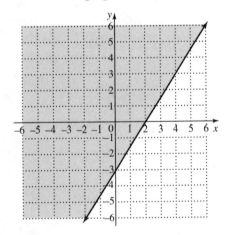

Now Try:

Graph $2x + 5y \leq -8$.

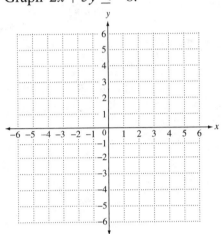

Graph $y \geq 3x$.

We graph $y = 3x$ using a solid line through $(0, 0)$, $(1, 3)$ and $(2, 6)$. Because $(0, 0)$ is on the line $y \geq 3x$, it cannot be used as a test point. Instead we choose a test point off the line, say $(3, 0)$.

$0 \overset{?}{\geq} 3(3)$

$0 \geq 9$ False

Because $0 \geq 9$ is false, shade the other region.

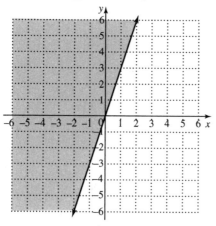

Graph $y \geq x$.

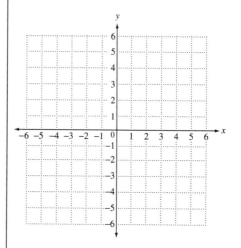

Practice Exercises

Graph each linear inequality.

10. $x - y < 5$

10.

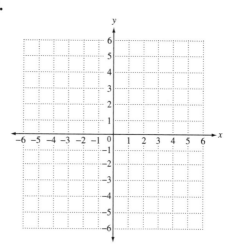

11. $2x + 3y \geq 6$

11.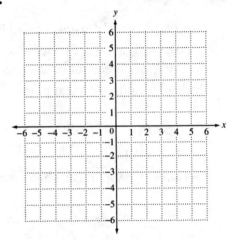

12. $x \leq 4y$

12.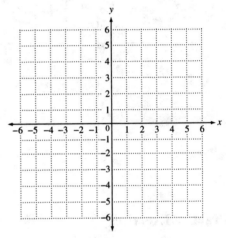

Chapter 11 Sequences and Series

Learning Objectives
Use inductive reasoning to analyze patterns.
Recognize and evaluate polynomial functions.

Key Terms

Use the vocabulary terms listed below to complete each statement in exercises 1–2.

inductive reasoning   **polynomial function of degree n**

1. A function defined by $f(x) = a_n x^n + a_{n-1} x^{n-1} + \cdots + a_1 x + a_0$, where $a_n \neq 0$ and n is a whole number is _____.

2. General conclusions drawn from specific observations is called

    _____.

Objective   Use inductive reasoning to analyze patterns.

Practice Exercises

Find the next number in each sequence. Describe the pattern in each sequence.

1.   4, 11, 18, 25, …

1. _____

2.   5, 10, 9, 14, 13, …

2. _____

3.   2, 6, 18, 54, 162, …

3. _____

Objective   Recognize and evaluate polynomial functions.

Video Examples

Review this example:

Let $f(x) = 6x^3 - 6x + 1$. Find $f(-3)$.

Substitute -3 for x.

$$f(-3) = 6(-3)^3 - 6(-3) + 1$$
$$= 6(-27) - 6(-3) + 1$$
$$= -162 + 18 + 1$$
$$= -143$$

Now Try:

Let $p(x) = -x^4 + 3x^2 - x + 7$.
Find $p(2)$.

_____

Practice Exercises

For each polynomial function, find (a) f(−2) and (b) f(3).

4. $f(x) = -x^2 - x - 5$

4. (a) _____

   (b) _____

5. $f(x) = 2x^2 + 3x - 5$

5. (a) _____

   (b) _____

6. $f(x) = 3x^4 - 5x^2$

6. (a) _____

   (b) _____

Chapter R REVIEW OF THE REAL NUMBER SYSTEM

R.1   Basic Concepts

Learning Objectives
1   Write sets using set notation.
2   Use number lines.
3   Know the common sets of numbers.
4   Find additive inverses.
5   Use absolute value.
6   Use inequality symbols.

Key Terms

Use the vocabulary terms listed below to complete each statement in exercises 1−15.

| set | elements | finite set | infinite set | empty set | variable |

set-builder notation   **number line** **coordinate** **graph**

additive inverse   **signed numbers**   **absolute value**

equation   **inequality**

1. A _____ is a line with a scale to indicate the set of real numbers.

2. _____ is used to describe a set of numbers without listing them.

3. The number that corresponds to a point on the number line is the _____ of that point.

4. A collection of objects is a _____.

5. If the number of elements in a set can be listed or counted and the counting process comes to an end, then the set is a(n) _____.

6. The set with no elements is called the _____.

7. A mathematical statement that two quantities are not equal is a(n) _____.

8. Positive and negative numbers are _____.

9. If the number of elements in a set cannot be listed or counted, then the set is a(n) _____.

10. A letter used to represent a number or a set of numbers is a(n) _____.

11. The _____ of a number is its distance from 0 on the number line.

12. The _____ of a number a is $-a$.

13. A mathematical statement that two quantities are equal is a(n) _____.

14. The point of the number line that corresponds to a number is its _____.

15. The _____ of a set are the numbers or objects that make up the set.

Objective 1   Write sets using set notation.

Video Examples

Review these examples for Objective 1:

1. List the elements in the set.

$\{x \mid x$ is a natural number less than $6\}$

The natural numbers less than 6 are 1, 2, 3, 4, 5. The set is {1, 2, 3, 4, 5}.

2. Use set-builder notation to describe each set.

$\{2, 4, 6, 8, 10\}$

One way to describe this set is $\{x \mid x$ is one of the first five even natural numbers$\}$.

Now Try:

1. List the elements in the set.

$\{x \mid x$ is a whole number less than $6\}$

_____

2. Use set-builder notation to describe each set.
$\{1, 2, 3, 4, 5\}$

_____

Objective 1 Practice Exercises

For extra help, see Examples 1–2 on page 2 of your text.

Write the set by listing its elements.

1. $\{y \mid y$ is a natural number divisible by $5\}$

1. _____

Write each set using set-builder notation. (More than one description is possible.)

2. $\{1,3,5,7,9\}$

2. _____

3. $\{7,8,9,10\}$

3. _____

Objective 2   Use number lines.

Objective 2 Practice Exercises

For extra help, see pages 2–4 of your text.

Graph the elements of each set on a number line.

4. $\left\{-\frac{1}{3}, 0, \frac{1}{3}, \frac{2}{3}\right\}$

4.

‹+++++++++++++++++›

5. $\left\{-4, -2.5, 0, \sqrt{9}\right\}$

5.

‹+++++++++++++++++›

6. $\left\{-4, -\frac{3}{2}, \frac{1}{2}, 5\right\}$

6.

‹+++++++++++++++++›

Objective 3   Know the common sets of numbers.

Video Examples

Review these examples for Objective 3:

3.   List the numbers in the following set that are elements of each set.

$$\left\{-6, -\sqrt{5}, -\frac{7}{40}, 0, 0.3, \frac{5}{9}, 1.\overline{75}, \sqrt{6}, 4\right\}$$

a. Integers

$-6, 0, 4$

b. Rational numbers

$-6, -\frac{7}{40}, 0, 0.3, \frac{5}{9}, 1.\overline{75}, 4$

c. Irrational numbers

$-\sqrt{5}, \sqrt{6}$

d. Real numbers

All are real numbers.

Now Try:

3.   List the numbers in the following set that are elements of each set.

$$\left\{-7, -\sqrt{3}, -\frac{8}{9}, 0, 0.4, \frac{3}{5}, 2.\overline{15}, \sqrt{7}, 3\right\}$$

a. Integers

_____

b. Rational numbers

_____

c. Irrational numbers

_____

d. Real numbers

_____

Objective 3 Practice Exercises

For extra help, see Examples 3–4 on page 5 of your text.

Which elements of each set are (a) natural numbers, (b) whole numbers, (c) integers, (d) rational numbers, (e) irrational numbers, (f) real numbers?

7. $\left\{-4, -\sqrt{2}, -\frac{1}{3}, 0, \frac{4}{5}, \sqrt{7}, 5\right\}$

7. a._____

   b._____

   c._____

   d._____

   e._____

   f._____

Decide if each statement is **true** *or* **false**.

8. All rational numbers are real numbers.

8. _____

9. Some irrational numbers are rational numbers.

9. _____

Objective 4   Find additive inverses.

Objective 4 Practice Exercises

For extra help, see pages 5–6 of your text.

Give the additive inverse of each number.

10. 0

10. _____

11. $-\frac{5}{2}$

11. _____

12. $\sqrt{2}$

12. _____

Objective 5   Use absolute value.

Video Examples

Review these examples for Objective 5:	Now Try:

Review these examples for Objective 5:

5. Find each value.

    a. $|-4|$

        $|-4| = -(-4) = 4$

    b. $-|7|$

        $-|7| = -(7) = -7$

    c. $-|-7|$

        $-|-7| = -(7) = -7$

    d. $|-4| + |3|$

        $|-4| + |3| = 4 + 3 = 7$

6. The table below shows the changes in population for five cities.

City	1980−1990	1990−2000	2000−2009
New York	250,925	685,714	383,603
Los Angeles	518,548	209,422	137,048
Chicago	−221,346	112,290	−44,748
Houston	35,415	323,078	304,295
Philadelphia	−102,633	−68,027	29,747

Source: factfinder.census.gov

Which city had the greatest change in population in which year?

New York from 1990−2000.

Now Try:

5. Find each value.

    a. $|-11|$

        _____

    b. $-|10|$

        _____

    c. $-|-10|$

        _____

    d. $|-6| + |3|$

        _____

6. Use the table shown at the left to determine which city has the least change in population in which year.

    _____

Objective 5 Practice Exercises

For extra help, see Examples 5–6 on page 7 of your text.

Find the value of each expression.

13. $\left|6\right| - \left|-1\right|$

14. $\left|-7\right| + \left|-8\right|$

13. _____

14. _____

Name: Date:

Instructor: Section:

Objective 6   Use inequality symbols.

Video Examples

Review these examples for Objective 6:

7.   Use a number line to compare –5 and –1, and to compare 0 and –1.

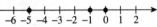

    –5 is located to the left of –1. For this reason, –5 < –1. Also, –1 > –5. From the same number line, –1 < 0, or 0 > –1.

8.   Decide whether the statement is true or false.

$$7 \cdot 3 \leq 6(4)$$

    True. $21 < 24$

Now Try:

7.   Insert < or > in the blank to make a true statement.

    $-5 \underline{\hphantom{xx}} -9$

    _____

8.   Decide whether the statement is true or false.

    $3 \cdot 8 > 4(6)$

    _____

Objective 6 Practice Exercises

For extra help, see Examples 7–8 on pages 8–9 of your text.

Identify each inequality as **true** *or* **false**.

15. $-3 < -5$ 15. _____

16. $\dfrac{3}{4} > \dfrac{2}{3}$ 16. _____

17. $-\left|7-4\right| \leq -4$ 17. _____

Chapter R REVIEW OF THE REAL NUMBER SYSTEM

R.2   Operations on Real Numbers

Learning Objectives
1   Add real numbers.
2   Subtract real numbers.
3   Find the distance between two points on a number line.
4   Multiply real numbers.
5   Find the reciprocals and divide real numbers.

Key Terms

Use the vocabulary terms listed below to complete each statement in exercises 1−5.

> **sum   difference   product   reciprocals   quotient**

1.   The answer to a multiplication problem is called the _____.

2.   Pairs of numbers whose product is 1 are called _____.

3.   The answer to a subtraction problem is called the _____.

4.   The answer to a division problem is called the _____.

5.   The answer to an addition problem is called the _____.

Objective 1   Add real numbers.

Video Examples

Review these examples for Objective 1:

1. Find the sum.

$$-13+(-9)$$

Because –13 and –9 have the same sign, add their absolute values.

$$-13+(-9)=-\left(|-13|+|-9|\right)$$
$$=-(22)$$
$$=-22$$

2. Find the sum.

$$5+(-3)$$

Subtract the absolute values, 5 and 3. Because 5 has the greater absolute value, the sum must be positive.

$$5+(-3)=5-3=2$$

Now Try:

1. Find the sum.

$$-17+(-4)$$

_____

2. Find the sum.

$$12+(-9)$$

_____

Objective 1 Practice Exercises

For extra help, see Examples 1–2 on pages 13–14 of your text.

Find each sum.

1. $-5.1 + (-7.3)$ 1. _____

2. $-16.32 + 2.27$ 2. _____

3. $\dfrac{2}{11} + \left(-\dfrac{2}{3}\right)$ 3. _____

Objective 2   Subtract real numbers.

Video Examples

Review these examples for Objective 2:

3. Find the difference.

$-13 - 5$

$-13 - 5 = -13 + (-5) = -18$

4. Perform the indicated operation.

$12 - (-4) - 8 - 15$

$$12 - (-4) - 8 - 15 = (12 + 4) - 8 - 15$$
$$= 16 - 8 - 15$$
$$= 8 - 15$$
$$= 8 + (-15)$$
$$= -7$$

Now Try:

3. Find the difference.

$-15 - 7$

_____

4. Perform the indicated operation.

$19 - (-9) - 8 - 7$

_____

Objective 2 Practice Exercises

For extra help, see Examples 3–4 on pages 15–16 of your text.

Find each difference.

4. $-3 - 7$ 4. _____

5. $-6.25 - (-2.47)$ 5. _____

6. $\dfrac{3}{5}-\left(-\dfrac{1}{3}\right)$

6. _____

Objective 3   Find the distance between two points on a number line.

Video Examples

Review this example for Objective 3:

5. Find the distance between the points –6 and 3.

Find the absolute value of the difference of the numbers, taken in either order.

$|-6-3|=|-9|=9,$ or $|3-(-6)|=|3+6|=9$

Now Try:

5. Find the distance between the points –7 and 5.

_____

Objective 3 Practice Exercises

For extra help, see Example 5 on page 16 of your text.

Find the distance between each pair of points.

7. –4 and –1

7. _____

8. 5 and –2

8. _____

9. –1 and 8

9. _____

Objective 4   Multiply real numbers.

Video Examples

Review these examples for Objective 4:

6. Find each product.

a. $-4(-6)$

The numbers have the same sign, so the product is positive.

$-4(-6)=24$

b. $\dfrac{3}{4}(-8)$

$\dfrac{3}{4}(-8)=-6$

Now Try:

6. Find each product.

a. $-7(-11)$

_____

b. $\dfrac{5}{6}(-12)$

_____

Objective 4 Practice Exercises

For extra help, see Example 6 on page 17 of your text.

Find each product.

10. $(-5)(7)$ 10. _____

11. $-1.2(-3.27)$ 11. _____

12. $\dfrac{13}{11}\left(-\dfrac{33}{26}\right)$ 12. _____

Objective 5   Find the reciprocals and divide real number.

Video Examples

Review these examples for Objective 5:

7.   Find each quotient.

   a. $\dfrac{-15}{5}$

   The numbers have opposite signs, so the quotient is negative.

   $\dfrac{-15}{5} = -3$

   b. $\dfrac{21}{-7}$

   $\dfrac{21}{-7} = -3$

   c. $\dfrac{0}{-1}$

   $\dfrac{0}{-1} = 0$

   d. $\dfrac{8}{0}$

   $\dfrac{8}{0}$ is undefined.

Now Try:

7.   Find each quotient.

   a. $\dfrac{-16}{8}$

   _____

   b. $\dfrac{27}{-9}$

   _____

   c. $\dfrac{0}{-3}$

   _____

   d. $\dfrac{-6}{0}$

   _____

Objective 5 Practice Exercises

For extra help, see Example 7 on page 19 of your text.

Give the reciprocal of each number.

13. -5 13. _____

14. $-\dfrac{11}{18}$ 14. _____

Divide where possible.

15. $\dfrac{-7}{0}$ 15. _____

11

Chapter R REVIEW OF THE REAL NUMBER SYSTEM

R.3   Exponents, Roots, and Order of Operations

Learning Objectives
Learning Objectives
1   Use exponents.
2   Find square roots.
3   Use the rules for order of operations.
4   Evaluate algebraic expressions for given values of variables.

Key Terms

Use the vocabulary terms listed below to complete each statement in exercises 1−6.

> **factors**   **exponent**   **base**   **exponential expression**
>
> **square root**   **algebraic expression**

1. A number written with an exponent is an _____.

2. The _____ is the number that is a repeated factor when written with an exponent.

3. A(n) _____ is a number that indicates how many times a factor is repeated.

4. A _____ of a number r is a number that can be multiplied by itself to obtain r.

5. A collection of numbers, variables, operation symbols, and grouping symbols is a(n) _____.

6. Two or more numbers whose product is a third number are _____ of that third number.

Objective 1   Use exponents.

Video Examples

Review these examples for Objective 1:
1. Write using exponents.

   a. $5 \cdot 5 \cdot 5 \cdot 5$

     $5 \cdot 5 \cdot 5 \cdot 5 = 5^4$

   b. $\dfrac{5}{6} \cdot \dfrac{5}{6} \cdot \dfrac{5}{6}$

     $\dfrac{5}{6} \cdot \dfrac{5}{6} \cdot \dfrac{5}{6} = \left(\dfrac{5}{6}\right)^3$

Now Try:
1. Write using exponents.

   a. $9 \cdot 9 \cdot 9$

   _____

   b. $\dfrac{3}{8} \cdot \dfrac{3}{8} \cdot \dfrac{3}{8} \cdot \dfrac{3}{8}$

   _____

c. $(-7)(-7)(-7)$

$$(-7)(-7)(-7) = (-7)^3$$

d. $x \cdot x \cdot x \cdot x \cdot x$

$$x \cdot x \cdot x \cdot x \cdot x = x^5$$

2. Evaluate.

a. 6^2

$$6^2 = 6 \cdot 6 = 36$$

b. $\left(\dfrac{3}{4}\right)^3$

$$\left(\dfrac{3}{4}\right)^3 = \dfrac{3}{4} \cdot \dfrac{3}{4} \cdot \dfrac{3}{4} = \dfrac{27}{64}$$

c. $(-3)^4$

$$(-3)^4 = (-3)(-3)(-3)(-3) = 81$$

d. $(-5)^3$

$$(-5)^3 = (-5)(-5)(-5) = -125$$

e. -3^4

-3^4 means $-3 \cdot 3 \cdot 3 \cdot 3$ which equals -81.

c. $(-10)(-10)(-10)(-10)$

_____

d. $z \cdot z \cdot z \cdot z \cdot z \cdot z$

_____

2. Evaluate.

a. 7^2

_____

b. $\left(\dfrac{2}{5}\right)^3$

_____

c. $(-4)^4$

_____

d. $(-2)^5$

_____

e. -5^4

_____

Objective 1 Practice Exercises

For extra help, see Examples 1–2 on pages 24–25 of your text.

Write the expression using exponents.

1. $b \cdot b \cdot b \cdot b \cdot b \cdot b \cdot b$

1. _____

Evaluate each expression.

2. $-(-4)^2$

2. _____

3. $(-2)^3$

3. _____

Objective 2   Find square roots.

Video Examples

Review these examples for Objective 2:

3. Find each square root that is a real number.

   a. $\sqrt{64}$

   $\sqrt{64} = 8$, since $8^2 = 64$.

   b. $-\sqrt{400}$

   $-\sqrt{400} = -20$, since the negative sign is outside the radical symbol.

   c. $\sqrt{\dfrac{4}{25}}$

   $\sqrt{\dfrac{4}{25}} = \dfrac{2}{5}$, since $\left(\dfrac{2}{5}\right)^2 = \dfrac{4}{25}$.

   d. $\sqrt{-400}$

   $\sqrt{-400}$ is not a real number since the negative sign is inside the radical symbol.

Now Try:

3. Find each square root that is a real number.

   a. $\sqrt{49}$

   _____

   b. $-\sqrt{900}$

   _____

   c. $\sqrt{\dfrac{16}{81}}$

   _____

   d. $\sqrt{-900}$

   _____

Objective 2 Practice Exercises

For extra help, see Example 3 on pages 25–26 of your text.

Find each square root. If it is not a real number, say so.

4. $-\sqrt{\dfrac{4}{49}}$

5. $\sqrt{10,000}$

6. $\sqrt{-81}$

4. _____

5. _____

6. _____

Objective 3   Use the rules for order of operations.

Video Examples

Review these examples for Objective 3:

4. Simplify $7 + 3 \cdot 4$.

   $7 + 3 \cdot 4 = 7 + 12$
   $\qquad\quad = 19$

Now Try:

4. Simplify $8 + 5 \cdot 6$.

   _____

5. Simplify $6 \cdot 5^2 + 3 - |-7 + 4|$

$$6 \cdot 5^2 + 3 - |-7 + 4| = 6 \cdot 5^2 + 3 - |-3|$$
$$= 6 \cdot 5^2 + 3 - 3$$
$$= 6 \cdot 25 + 3 - 3$$
$$= 150 + 3 - 3$$
$$= 153 - 3$$
$$= 150$$

6. Simplify $\dfrac{9 + 5^3}{8\sqrt{16} - 2^5}$.

$$\frac{9 + 5^3}{8\sqrt{16} - 2^5} = \frac{9 + 125}{8 \cdot 4 - 32}$$
$$= \frac{9 + 125}{32 - 32}$$
$$= \frac{134}{0}$$

Because division by 0 is undefined, the given expression is undefined.

5. Simplify $8 \cdot 2^2 + 6 - |-9 + 5|$

_____

6. Simplify $\dfrac{-5(9) + 4^2}{2\sqrt{81} - \dfrac{1}{3}(54)}$.

_____

Objective 3 Practice Exercises

For extra help, see Examples 4–6 on pages 26–27 of your text.

Simplify each expression.

7. $4^3 \div 2^5 + 3\sqrt{36}$

7. _____

8. $-4(-3)^3 - 5(8 - 4)$

8. _____

9. $\dfrac{6(-3) + (-5)^2(-4)}{11 - 3^2}$

9. _____

Objective 4   **Evaluate algebraic expressions for given values of variables.**

Video Examples

Review these examples for Objective 4:

7. Evaluate each expression for $m = -3$, $n = 6$, $p = -8$, and $q = 16$.

   a. $12m - 8n$

Substitute -3 for m and 6 for n.
$$12m - 8n = 12(-3) - 8(6)$$
$$= -36 - 48$$
$$= -84$$

   b. $-5n^2 - m^2\left(\sqrt{q}\right)$

$$-5n^2 - m^2\left(\sqrt{q}\right) = -5(6)^2 - (-3)^2\left(\sqrt{16}\right)$$
$$= -5(36) - (9)(4)$$
$$= -180 - 36$$
$$= -216$$

Now Try:

7. Evaluate each expression for $m = -5$, $n = 8$, $p = -10$, and $q = 36$.

   a. $7m - 5n$

_____

   b. $-6m^2 + n^2\left(\sqrt{q}\right)$

_____

Objective 4 Practice Exercises

For extra help, see Example 7 on page 28 of your text.

Evaluate each expression if a = –1, b = 3, and c = –5.

10. $2a^3 - b^2$ 10. _____

11. $\dfrac{2a + 4b^2}{3c - 2a}$ 11. _____

12. $\dfrac{6c - b}{7a^3 - 4c}$ 12. _____

Chapter R REVIEW OF THE REAL NUMBER SYSTEM

R.4   Properties of Real Numbers

Learning Objectives
1   Use the distributive property.
2   Use the identity properties.
3   Use the inverse properties.
4   Use the commutative and associative properties.

Key Terms

Use the vocabulary terms listed below to complete each statement in exercises 1−4.

term   coefficient   like terms   combining like terms

1. A _____ is the numerical factor of a term.

2. A number, a variable, or a product or quotient of a number and one or more variables raised to powers is called a _____.

3. Terms with exactly the same variables, including the same exponents, are called _____.

4. Adding or subtracting like terms by using the properties of real numbers is called _____.

Objective 1   Use the distributive property.

Video Examples

Review these examples for Objective 1:

1. Use the distributive property to rewrite each expression.

   a. $-5(8+m)$

   $-5(8+m) = -5(8) - 5(m)$
   $= -40 - 5m$

   b. $6r - 11r$

   $6r - 11r = 6r + (-11r)$
   $= [6 + (-11)]r$
   $= -5r$

Now Try:

1. Use the distributive property to rewrite each expression.

   a. $-4(3+q)$

   _____

   b. $9r - 15r$

   _____

Objective 1 Practice Exercises

For extra help, see Example 1 on page 33 of your text.

Use the distributive property to rewrite each expression.

   1. $-3(z-7)$ **1.** _____

   2. $6y-15y$ **2.** _____

Use the distributive property to calculate the expression mentally.

   3. $24\cdot13+26\cdot13$ **3.** _____

Objective 2   **Use the identity properties.**

Video Examples

Review these examples for Objective 2:

2.   Simplify each expression.

      a. $17m+m$

$$17m+m=17m+1m$$
$$=(17+1)m$$
$$=18m$$

      b. $-(a-9z)$

$$-(a-9z)=-1(a-9z)$$
$$=-1(a)+(-1)(-9z)$$
$$=-a+9z$$

Now Try:

2.   Simplify each expression.

      a. $29q+q$

           _____

      b. $-(4-5c)$

           _____

Objective 2 Practice Exercises

For extra help, see Example 2 on page 34 of your text.

Complete each statement.

   4. $-6+0=$ _____   **4.** _____

   5.   _____ $+0=-2.5$ **5.** _____

   6. $-\dfrac{7}{3}\cdot1=$ _____   **6.** _____

Objective 3   Use the inverse properties.

Objective 3 Practice Exercises

For extra help, see pages 34–35 of your text.

Complete each statement.

7. $5 \cdot$ _____ $= 1$ 7. _____

8. $-\dfrac{3}{4}\left(-\dfrac{4}{3}\right) =$ _____   8. _____

9. $\dfrac{1}{7} +$ _____ $= 0$ 9. _____

Objective 4   Use the commutative and associative properties.

Video Examples

Review these examples for Objective 4:

3.   Simplify $-6x + 9x - 4 + 8x - 6$.

$$-6x + 9x - 4 + 8x - 6 = (-6x + 9x) - 4 + 8x - 6$$
$$= (-6 + 9)x - 4 + 8x - 6$$
$$= 3x - 4 + 8x - 6$$
$$= [3x + (-4 + 8x)] - 6$$
$$= [3x + (8x - 4)] - 6$$
$$= [(3x + 8x) - 4] - 6$$
$$= (11x - 4) - 6$$
$$= 11x - 4 - 6$$
$$= 11x - 10$$

4.   Simplify each expression.

   a. $7x + 5 - 6(x + 3) + 14$

$$7x + 5 - 6(x + 3) + 14 = 7x + 5 - 6x - 18 + 14$$
$$= 7x - 6x + 5 - 18 + 14$$
$$= x + 1$$

Now Try:

3.   Simplify $-7x + 3x - 9 + 2x - 7$.

_____

4.   Simplify each expression.

   a. $4y + 5 - 3(y + 3) - 2$

_____

b. $(4x)(7)(z)$

$$(4x)(7)(z) = [(4x)(7)]z$$
$$= [4(x \cdot 7)]z$$
$$= [4(7x)]z$$
$$= [(4 \cdot 7)x]z$$
$$= [28x]z$$
$$= 28(xz)$$
$$= 28xz$$

b. $(6x)(4)y$

_____

Objective 4 Practice Exercises

For extra help, see Examples 3–4 on pages 36–37 of your text.

Simplify each expression.

10. $-(2w-7)+11+3(4w-6)-5w$

10. _____

11. $2(x-3)-7-6(2x-5)+7x$

11. _____

12. $8-2(4d-1)+3(d-6)+2d$

12. _____

Chapter 1 LINEAR EQUATIONS, INEQUALITIES, AND APPLICATIONS

1.1 Linear Equations in One Variable

Learning Objectives
1   Distinguish between expressions and equations.
2   Identify linear equations.
3   Solve linear equations by using the addition and multiplication properties of equality.
4   Solve linear equations by using the distributive property.
5   Solve linear equations with fractions or decimals.
6   Identify conditional equations, contradictions, and identities.

Key Terms

Use the vocabulary terms listed below to complete each statement in exercises 1−7.

linear (first-degree) equation in one variable   **solution**

solution set   **equivalent equations**

conditional equation   **identity**   **contradiction**

1.   Equations that have exactly the same solution sets are called

_____.

2.   An equation that can be written in the form $Ax + B = C$, where A, B, and C are real numbers and $A \neq 0$, is called a _____.

3.   The set of all numbers that satisfy an equation is called its

_____.

4.   An equation with no solution is called a(n) _____.

5.   A(n) _____ is an equation that is true for some values of the variable and false for other values.

6.   An equation that is true for all values of the variable is called a(n)

_____.

7.   A _____ of an equation is a number that makes the equation true when substituted for the variable.

Objective 1   Distinguish between expressions and equations.

Objective 1 Practice Exercises

For extra help, see Example 1 on page 44 of your text.

Decide whether each of the following is an **expression** *or an* **equation**.

1. $13 - k = 11$

2. $\dfrac{1}{3}q - \dfrac{5}{3} = \dfrac{7}{3}$

3. $6k - 4$

1. _____

2. _____

3. _____

Objective 2   Identify linear equations.

Objective 2 Practice Exercises

For extra help, see pages 44–45 of your text.

Decide whether each of the following is a **linear equation** *or a* **nonlinear equation**.

4. $13 - k = 11$

5. $\dfrac{1}{q} - \dfrac{5}{3} = \dfrac{7}{3}$

6. $x^2 + 2y = 9$

4. _____

5. _____

6. _____

Objective 3   Solve linear equations by using the addition and multiplication properties of equality.

Video Examples

Review this example for Objective 3:
2.   Solve $5x - 3x - 6 = 9 + 4x + 3$.

$$5x - 3x - 6 = 9 + 4x + 3$$
$$2x - 6 = 12 + 4x$$
$$2x - 6 - 4x = 12 + 4x - 4x$$
$$-2x - 6 = 12$$
$$-2x - 6 + 6 = 12 + 6$$
$$-2x = 18$$
$$\frac{-2x}{-2} = \frac{18}{-2}$$
$$x = -9$$

Now Try:
2.   Solve $6x - 3x - 7 = 8 + x - 5$.

_____

Check   Substitute –9 for x in the original equation.

$$5x - 3x - 6 = 9 + 4x + 3$$
$$5(-9) - 3(-9) - 6 = 9 + 4(-9) + 3$$
$$-45 + 27 - 6 = 9 - 36 + 3$$
$$-24 = -24$$

Objective 3 Practice Exercises

For extra help, see Example 2 on pages 45–46 of your text.

Solve and check each equation.

7. $9r - 4r + 8r - 6 = 10r - 11 + 2r$

7. _____

8. $9x - 4 + 6x - 1 = 12 - 2x$

8. _____

9. $18 - 3y + 11 = 7y + 6y - 27 - 9y$

9. _____

Objective 4   Solve linear equations by using the distributive property.

Video Examples

Review this example for Objective 4:

3.   Solve $5(k - 4) - k = k - 2$.

Step 1 Since there are no fractions or decimals, Step 1 does not apply.

Step 2 $5(k - 4) - k = k - 2$
$$5(k) + 5(-4) - k = k - 2$$
$$5k - 20 - k = k - 2$$
$$4k - 20 = k - 2$$

Now Try:

3.   Solve $9(k - 2) - k = k + 10$.

_____

Step 3 $4k - 20 - k = k - 2 - k$

$3k - 20 = -2$

$3k - 20 + 20 = -2 + 20$

$3k = 18$

Step 4 $\dfrac{3k}{3} = \dfrac{18}{3}$

$k = 6$

Step 5 Check by substituting 6 for k in the original equation.

$5(k - 4) - k = k - 2$

$5(6 - 4) - 6 \overset{?}{=} 6 - 2$

$5(2) - 6 \overset{?}{=} 4$

$10 - 6 \overset{?}{=} 4$

$4 = 4$ True

The solution, 6, checks, so the solution set is $\{6\}$.

Objective 4 Practice Exercises

For extra help, see Examples 3–4 on pages 46–47 of your text.

Solve and check each equation.

10. $5(2p + 1) - (p + 3) = 7$ **10.** _____

11. $-[p - (4p + 2)] = 3 + (4p + 7)$ **11.** _____

12. $-9w - (4 + 3w) = -(2w - 1) - 5$ **12.** _____

Objective 5   Solve linear equations with fractions or decimals.

Video Examples

Review this example for Objective 5:

5.   Solve $\dfrac{x+8}{3}+\dfrac{3x-5}{6}=6$.

$$\dfrac{x+8}{3}+\dfrac{3x-5}{6}=6$$

Step 1 $6\left(\dfrac{x+8}{3}+\dfrac{3x-5}{6}\right)=6(6)$

$$6\left(\dfrac{x+8}{3}\right)+6\left(\dfrac{3x-5}{6}\right)=6(6)$$

$$2(x+8)+3x-5=36$$

$$2x+16+3x-5=36$$

$$5x+11=36$$

Step 2 $5x+11-11=36-11$

Step 3 $\dfrac{5x}{5}=\dfrac{25}{5}$

$$x=5$$

Step 4 $\dfrac{x+8}{3}+\dfrac{3x-5}{6}=6$

$$\dfrac{5+8}{3}+\dfrac{3(5)-5}{6}\overset{?}{=}6$$

$$\dfrac{13}{3}+\dfrac{15-5}{6}\overset{?}{=}6$$

$$\dfrac{13}{3}+\dfrac{10}{6}\overset{?}{=}6$$

$$\dfrac{13}{3}+\dfrac{5}{3}\overset{?}{=}6$$

$$6=6\quad\text{True}$$

The solution set is $\{5\}$.

Now Try:

5.   Solve $\dfrac{2x+7}{5}+\dfrac{3x-9}{10}=-3$.

_____

Objective 5 Practice Exercises

For extra help, see Examples 5–6 on pages 48–49 of your text.

Solve and check each equation.

13. $\dfrac{x-5}{2}-\dfrac{x+6}{3}=-4$

13.   _____

14. $\dfrac{2x+5}{5} - \dfrac{3x+1}{2} = \dfrac{7-x}{2}$

14. _____

15. $0.35(140) + 0.15w = 0.05(w + 1100)$

15. _____

Objective 6   Identify conditional equations, contradictions, and identities.

Video Examples

Review these examples for Objective 6:

7. Solve each equation. Decide whether it is a *conditional equation*, an *identity*, or a *contradiction*.

   a. $6(2x+7) - 2 = 4(x+10)$

   $$6(2x+7) - 2 = 4(x+10)$$
   $$12x + 42 - 2 = 4x + 40$$
   $$12x + 40 = 4x + 40$$
   $$12x - 4x + 40 - 40 = 4x - 4x + 40 - 40$$
   $$\frac{8x}{8} = \frac{0}{8}$$
   $$x = 0$$

   The solution 0 checks, so the solution set is $\{0\}$. Since the solution set has only one element, $6(2x+7) - 2 = 4(x+10)$ is a conditional equation.

   b. $9x - 11 = 9(x-2)$

   $$9x - 11 = 9(x-2)$$
   $$9x - 11 = 9x - 18$$
   $$9x - 11 - 9x = 9x - 18 - 9x$$
   $$-11 = -18 \quad \text{False}$$

   Since the result is false, the equation has no solution. The solution set is $\varnothing$, so the equation $9x - 11 = 9(x-2)$ is a contradiction.

Now Try:

7. Solve each equation. Decide whether it is a *conditional equation*, an *identity*, or a *contradiction*.

   a. $5x - 19 = 5(7x - 4) + 1$

   _____

   b. $10x - 4 = 5(2x + 2) - 3$

   _____

c. $8x + 5 = 8(x+1) - 3$ **c.** $6x - 13 = 6(x-3) + 5$

$$8x + 5 = 8(x+1) - 3$$
$$8x + 5 = 8x + 8 - 3$$
$$8x + 5 = 8x + 5$$
$$8x + 5 - 8x - 5 = 8x + 5 - 8x - 5$$
$$0 = 0 \quad \text{True}$$

The final line gives a true statement, which indicates that the solution set is {all real numbers}. The equation $8x + 5 = 8(x+1) - 3$ is an identity.

Objective 6 Practice Exercises

For extra help, see Example 7 on pages 49–50 of your text.

Decide whether each equation is a **conditional equation**, *an* **identity**, *or a* **contradiction**. *Give the solution set.*

16. $7(2 - 5b) - 32 = 10b - 3(6 + 15b)$ **16.** _____

17. $7(3 - 4q) - 10(q - 2) = 19(5 - 2q)$ **17.** _____

18. $13p - 9(3 - 2p) = 3(10p - 9) + 1$ **18.** _____

Chapter 1 LINEAR EQUATIONS, INEQUALITIES, AND APPLICATIONS

1.2   Formulas and Percent

Learning Objectives

1   Solve a formula for a specified variable.
2   Solve applied problems using formulas.
3   Solve percent problems.
4   Solve problems involving percent increase or decrease.
5   Solve problems from the health care industry.

Key Terms

Use the vocabulary terms listed below to complete each statement in exercises 1–3.

mathematical model   **formula**   **percent**

1. An equation in which variables are used to describe a relationship is a

   _____.

2. A _____ is an equation or inequality that describes a real situation.

3. _____ means "one per hundred."

Objective 1   Solve a formula for a specified variable.

Video Examples

Review these examples for Objective 1:

1. Solve the formula $V = LWH$ for L.

$$V = LWH$$
$$V = L(WH)$$
$$\frac{V}{WH} = \frac{L(WH)}{WH}$$
$$\frac{V}{WH} = L$$

2. Solve $F = \frac{9}{5}C + 32$, for C.

Step 1 Multiply by 5 to clear fractions.

$$5(F) = 5\left(\frac{9}{5}C + 32\right)$$
$$5F = 5\left(\frac{9}{5}C\right) + 5(32)$$
$$5F = 9C + 160$$

Step 2 $5F - 160 = 9C + 160 - 160$

$$5F - 160 = 9C$$

Now Try:

1. Solve the formula $A = LW$ for L.

   _____

2. Solve $V = \frac{1}{3}Bh$, for h.

   _____

$$Step \ 3 \quad \frac{5F-160}{9} = \frac{9C}{9}$$

$$\frac{5F-160}{9} = C$$

$$C = \frac{5F-160}{9}$$

$$C = \frac{5F}{9} - \frac{160}{9}$$

Objective 1 Practice Exercises

For extra help, see Examples 1–4 on pages 54–56 of your text.

Solve each formula or equation for the specified variable.

1. $V = \frac{1}{3}Bh$ for B

    1. _____

2. $r = \frac{7}{x+5}$ for x

    2. _____

3. $\frac{2+x}{3} = \frac{y}{6}$ for y

    3. _____

Objective 2   Solve applied problems using formulas.

Video Examples

Review this example for Objective 2:

5. It takes Grace $\frac{2}{3}$ hr to drive 30 miles. What is her average rate?

    We use the formula $d = rt$ solving for r.

    $$\frac{d}{t} = \frac{rt}{t}$$

    $$\frac{d}{t} = r, \text{ or } r = \frac{d}{t}$$

    Substitute the values for d and t.

    $$r = \frac{30}{\frac{2}{3}} = 30 \cdot \frac{3}{2} = 45$$

    Her average rate is 45 mph.

Now Try:

5. It takes Tyler $\frac{1}{4}$ hr to drive 15 miles. What is his average rate?

    _____

Objective 2 Practice Exercises

For extra help, see Example 5 on pages 56–57 of your text.

Solve each problem using the appropriate formula.

4. A cord of wood contains 128 cubic feet of wood. A   4. _____
    stack of wood is 4 feet high and 8 feet long. How
    wide must it be to contain a cord?

5. If $700 earns $112 simple interest in 2 years, find   5. _____
    the rate of interest.

6. The Fahrenheit temperature is 104°. Find the Celsius   6. _____
    temperature.

Objective 3   Solve percent problems.

Video Examples

Review these examples for Objective 3: | **Now Try:**
6. | 6.

a. An alcohol and water mixture measures 45 | a. A certificate of deposit pays
liters. The mixture contains 9 liters of alcohol. | $117 simple interest in one year
What percent of the mixture is alcohol? | on a principal of $4500. What
                                                    | interest rate is being paid on this
Let x represent the percent of alcohol in the | deposit?
mixture.

$$x = \frac{9}{45} \quad \begin{array}{l} \leftarrow \text{part} \\ \leftarrow \text{whole} \end{array}$$

$x = 0.20$ or 20%
The mixture is 20% alcohol.   _____

b. The purchase price of a new car is $12,500. In order to finance the car a purchaser is required to make a minimum down payment of 20% of the purchase price. What is the minimum down payment required?

Let x represent the amount of down payment.

$$\frac{x}{12,500} = 0.20$$

$$12,500 \cdot \frac{x}{12,500} = 12,500(0.20)$$

$$x = 2500$$

The down payment is $2500.

b. A salesperson earned $33,250 on annual sales of $950,000. What is her rate of commission?

_____

7. The graph shows spending by department at a college. The college's total budget is $280 million. How much was budgeted for the business department?

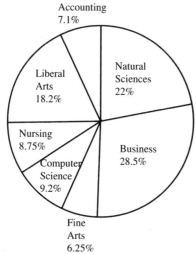

Accounting
7.1%

Liberal Arts
18.2%

Natural Sciences
22%

Nursing
8.75%

Computer Science
9.2%

Business
28.5%

Fine Arts
6.25%

7. Refer to the graph at the left. If the college's total budget is $330 million, how much is budgeted for fine arts?

_____

Since 28.5% was budgeted for the business department, let x = this amount in millions of dollars. 28.5% = 0.285.

$$\frac{x}{280} = 0.285$$

$$x = 0.285(280)$$

$$x = 79.8$$

$79.8 million was budgeted for the business department.

Objective 3 Practice Exercises

For extra help, see Examples 6–7 on page 58 of your text.

Solve each problem.

7. Twelve percent of a college student body has a grade point average of 3.0 or better. If there are 1250 students enrolled in the college, how many have a grade point average of less than 3.0?

7. _____

8. In a class of freshmen and sophomores only, there are 85 students. If 34 are freshmen, what percent of the class are sophomores?

8. _____

9. At a large university, 40% of the student body is from out of state. If the total enrollment at the university is 25,000 students, how many are from in-state?

9. _____

Objective 4   Solve problems involving percent increase or decrease.

Video Examples

Review these examples for Objective 4:

8. Use the percent change equation in each application.

    a. Over the last three years, Calvin's salary has increased from $2700 per month to $3200. What is the percent increase?

    Let *x* represent the percent increase.

    $$\text{percent increase} = \frac{\text{amount of increase}}{\text{original amount}}$$

    $$x = \frac{3200 - 2700}{2700}$$

    $$x = \frac{500}{2700}$$

    $$x \approx 0.185, \text{ or } 18.5\%$$

    Calvin's salary increased by approximately 18.5%.

Now Try:

8. Use the percent change equation in each application.

    a. Students at Withrow's College were charged $1560 for tuition this semester. If the tuition was $1480 last semester, find the percent of increase.

    _____

b. The number of days employees of Prodex Manufacturing Company were absent from their jobs decreased from 96 days last month to 72 days this month. Find the percent of decrease.

Let x represent the percent decrease.

$$\text{percent decrease} = \frac{\text{amount of decrease}}{\text{original amount}}$$

$$x = \frac{96 - 72}{96}$$

$$x = \frac{24}{96}$$

$$x = 0.25, \text{ or } 25\%$$

The absentee rate decreased by 25%.

b. During a sale, the price of a futon was cut from $1250 to $999. Find the percent of decrease in price.

_____

Objective 4 Practice Exercises

For extra help, see Example 8 on page 59 of your text.

Solve each problem. Round money answer to the nearest cent and percent answer to the nearest tenth, if necessary.

10. The sale price of a piece of furniture is $1020. This represents 15% off the regular price. What is the regular price?

10. _____

11. Mark Johnson invested $2500 in stock one year ago. During the year the stock increased in value by $162.50. What interest rate has Mark's investment earned?

11. _____

12. Tracy works as a massage therapist. During July, she cut her price on massages from $54 to $45.50. By what percent did she decrease the price?

12. _____

Objective 5   Solve problems from the health care industry.

Video Examples

Review these examples for Objective 5:

9.   If a child weighs k kilograms, then the child's body surface area S in square meters (m^2) is determined by the formula
$$S = \frac{4k+7}{k+90}.$$
Use the formula to find the body surface area, to the nearest hundredth, of a child who weighs 30 lb. Use $k = 13.608$.

Use the value for k in the formula.
$$S = \frac{4(13.608)+7}{13.608+90}$$
$$S = 0.59$$
The child's body surface area is 0.59 m^2.

10.   If D represents the usual adult dose of a medication in milligrams, the corresponding child's dose in milligrams is calculated by using the formula
$$C = \frac{\text{body surface area in square meters}}{1.7} \times D$$
Determine the appropriate dose, to the nearest unit, for a child whose body surface area is 0.67 m^2, if the usual adult dose is 100 mg.

Use the values for $S = 0.67$ m^2 and $D = 100$ mg in the formula.
$$C = \frac{0.67}{1.7} \times 100$$
$$C = 39$$
The child's dose area is 39 mg.

Now Try:

9.   If a child weighs k kilograms, then the child's body surface area S in square meters (m^2) is determined by the formula
$$S = \frac{4k+7}{k+90}.$$
Use the formula to find the body surface area, to the nearest hundredth, of a child who weighs 35 lb. Use $k = 15.876$.

_____

10.   If D represents the usual adult dose of a medication in milligrams, the corresponding child's dose in milligrams is calculated by using the formula
$$C = \frac{\text{body surface area in } m^2}{1.7} \times D$$
Determine the appropriate dose, to the nearest unit, for a child whose body surface area is 0.59 m^2, if the usual adult dose is 50 mg.

_____

Objective 5 Practice Exercises

For extra help, see Examples 9–10 on page 60 of your text.

Solve each problem.

13.   If a child weighs k kilograms, then the child's body   **13.** _____
surface area S in square meters (m^2) is determined by
the formula

$$S = \frac{4k + 7}{k + 90}.$$

Use the formula to find the body surface area, to the
nearest hundredth, of a child who weighs 30 kg.

14.   If D represents the usual adult dose of a medication   **14.** _____
in milligrams, the corresponding child's dose in
milligrams is calculated by using the formula

$$C = \frac{\text{body surface area in square meters}}{1.7} \times D$$

Determine the appropriate dose, to the nearest unit,
for a child whose body surface area is
0.87 m^2, if the usual adult dose is 100 mg.

Name: Date:
Instructor: Section:

Chapter 1 LINEAR EQUATIONS, INEQUALITIES, AND APPLICATIONS

1.3   Applications of Linear Equations

Learning Objectives
1   Translate from words to mathematical expressions.
2   Write equations from given information.
3   Distinguish between simplifying expressions and solving equations.
4   Use the six steps in solving an applied problem.
5   Solve percent problems.
6   Solve investment problems.
7   Solve mixture problems.

Key Terms

Use the vocabulary terms listed below to complete each statement in exercises 1−4.

sum	of	product	increased by
quotient	less than	per	double
decreased by	more than	difference	ratio

1. _____, _____, and

    _____ are words that mean addition.

2. _____, _____, and

    _____ are words that mean multiplication.

3. _____, _____, and

    _____ are words that mean division.

4. _____, _____, and

    _____ are words that mean subtraction.

Objective 1   Translate from words to mathematical expressions.

For extra help, see pages 66–67 of your text.

Translate each verbal phrase into a mathematical expression. Use x to represent the unknown number.

1. The product of a number and 7 subtracted from 13   1. _____

2. −7 increased by 4 times a number   2. _____

3. The quotient of 4 more than a number and 9   3. _____

Objective 2   Write equations from given information.

Video Examples

Review these examples for Objective 2:

1. Translate each verbal sentence into an equation.

   a. If the product of a number and 24 is decreased by 9, the result is 63.

   $$24x - 9 = 63$$

   b. The quotient of a number and the number minus 6 is 32.

   $$\frac{x}{x-6} = 32$$

Now Try:

1. Translate each verbal sentence into an equation.
   a. If the product of a number and 3 is decreased by 100, the result is 197.

   _____

   b. The quotient of a number and the number plus 5 is 23.

   _____

Objective 2 Practice Exercises

For extra help, see Example 1 on page 67 of your text.

Use the variable x for the unknown, and write an equation representing the verbal sentence. Do not solve.

4. 18 minus a number is equal to the number times 4.

   4. _____

5. The ratio of a number and the difference between the number and 3 is 17.

   5. _____

6. Twice a number is 3 times the sum of the number and 7.

   6. _____

Objective 3   Distinguish between simplifying expressions and solving equations.

Video Examples

Review these examples for Objective 3:

2. Decide whether each is an *expression* or an *equation*. Simplify the expression, and solve the equation.

   a. $5(7+x) - 3x + 11$

   This is an expression.
   $$5(7+x) - 3x + 11$$
   $$= 35 + 5x - 3x + 11$$
   $$= 2x + 46$$

Now Try:

2. Decide whether each is an *expression* or an *equation*. Simplify the expression, and solve the equation.
   a. $9(x-5) - 8x = 20$

   _____

b. $6(4+x)-5x+8=-4$

b. $20x-4(x-7)+3$

This is an equation.
$$6(4+x)-5x+8=-4$$
$$24+6x-5x+8=-4$$
$$32+x=-4$$
$$x=-36$$
The solution set is $\{-36\}$.

_____

Objective 3 Practice Exercises

For extra help, see Example 2 on page 67 of your text.

Decide whether each is an expression or an equation. Simplify the expression, and solve the equation.

7. $3(z-9)+8(2z-4)$

7. _____

8. $3(z-9)=8(2z-4)$

8. _____

9. $\dfrac{k}{3}-\dfrac{k+5}{4}=12$

9. _____

Objective 4   Use the six steps in solving an applied problem.

Video Examples

Review this example for Objective 4:

3. The width of a rectangle is $\frac{1}{4}$ of the difference between the length and 1 meter. If the perimeter of the rectangle is 62 meters, find the length and width.

Step 1 Read the problem. Find the length and width.

Step 2 Assign a variable. Let $L=$ length.

Then $\frac{1}{4}(L-1)=$ width.

Step 3 Write an equation. Use the formula for the perimeter of a rectangle.
$$P=2L+2W$$
$$62=2L+2\left[\frac{1}{4}(L-1)\right]$$

Now Try:

3. In a triangle, the shortest side is 3 meters less than the longest side. The length of the third side is 11 meters less than twice the length of the longest side. If the perimeter is 22 meters, find the lengths of the three sides of the triangle.

_____

Step 4 Solve.

$$62 = 2L + \frac{1}{2}(L-1)$$

$$2(62) = 2\left(2L + \frac{1}{2}(L-1)\right)$$

$$124 = 4L + L - 1$$

$$124 = 5L - 1$$

$$125 = 5L$$

$$\frac{125}{5} = \frac{5L}{5}$$

$$25 = L$$

Step 5 State the answer. The length is 25 m and the width is $\frac{1}{4}(25-1) = 6$ m.

Step 6 Check. The perimeter is $2(25) + 2(6) = 62$ m.

Objective 4 Practice Exercises

For extra help, see Examples 3–4 on pages 68–69 of your text.

Solve each problem.

10. Mrs. Henry has 36 feet of fencing, which she plans to install to form a square-shaped yard for her dog. If the back of her house becomes one side of the square with the other sides requiring fencing, how long should each side be?

    10. _____

11. In the controversial 2000 presidential election, George W. Bush and Al Gore received a total of 537 electoral votes. Bush received five more votes than Gore. How many votes did each candidate receive?

    11. _____

    _____

12. The sum of two numbers is 41. The larger number is 1 more than 4 times the smaller number. What are these numbers?

    12. _____

    _____

Objective 5   Solve percent problems.

Video Examples

Review this example for Objective 5:

5. A merchant in a small store took in $508.80 in one day. This included merchandise sales and the 6% sales tax. Find the amount of merchandise sold.

    Step 1 Read the problem. Find the amount sold.

    Step 2 Assign a variable. Let x = the amount sold.

    Step 3 Write an equation.
    $$x + 0.06x = 508.80$$

    Step 4 Solve.
    $$1.06x = 508.80$$
    $$\frac{1.06x}{1.06} = \frac{508.80}{1.06}$$
    $$x = 480$$

    Step 5 State the answer. The merchant sold $480 of merchandise.

    Step 6 Check. $480 + 0.06(480) = 508.80$
    The answer checks.

Now Try:

5. Jim wishes to make $235,000 selling his home. If his realtor must earn a 6% commission, for what price should Jim sell his home?

    _____

Objective 5 Practice Exercises

For extra help, see Example 5 on page 70 of your text.

Solve each problem.

13. A savings account earns interest at an annual rate of 4%. At the end of one year Sheri would like to have a total of $5200 in the account. How much should she deposit now?

13. _____

14. At the end of a day, a storekeeper had $642 in the cash registers, counting both the sale of goods and the sales tax of 7%. Find the amount of the tax.

14. _____

15. A certificate of deposit earns $2\frac{3}{4}\%$ annually. If $6200 is invested, what is the certificate of deposit worth in one year?

15. _____

Objective 6   Solve investment problems.

Video Examples

Review this example for Objective 6:

6. Larry invested some money at 8% simple interest and $700 less than this amount at 7%. His total annual income from the interest was $584. How much was invested at each rate?

Step 1 Read the problem. Find the two amounts.

Step 2 Assign a variable.
Let x = the amount at 8%.
Let $x - 700$ = the amount at 7%.

Step 3 Write an equation.
$$0.08x + 0.07(x - 700) = 584$$

Step 4 Solve.
$$0.08x + 0.07x - 49 = 584$$
$$0.15x - 49 = 584$$
$$0.15x = 633$$
$$x = 4220$$

Step 5 State the answer. Larry must invest $4220 at 8%, and $4220 - 700 = \$3520$ at 7%.

Step 6 Check. The sum of the two amounts should be $584.
$$0.08(4220) + 0.07(3520) = 584$$

Now Try:

6. Tracy invested some money at 5% and $300 more than twice this amount at 7%. Her total annual income from the two investments is $325. How much is invested at each rate?

_____

Objective 6 Practice Exercises

For extra help, see Example 6 on page 71 of your text.

Solve each problem.

16. Louisa invested $16,000 in bonds paying 7% simple   16. _____
     interest. How much additional money should she
     invest at 4% simple interest so that the average
     return on the two investments is 6%?

17. Desiree invested some money at 9% and $100 less   17. 7%: _____
     than three times that amount at 7%. Her total annual
     interest was $83. How much did she invest at each   9%: _____
     rate?

18. Jacob has $48,000 invested in stocks paying 6%.   18. _____
     How much additional money should he invest in
     certificates of deposit paying 2.5% so that the
     average return on the two investments is 4%?

Objective 7   Solve mixture problems.

Video Examples

Review these examples for Objective 7:	Now Try:
7. How many gallons of a 20% alcohol solution must be mixed with 15 gallons of a 12% alcohol solution to obtain a 14% alcohol solution?	7. How many ounces of a 35% solution of acid must be mixed with a 60% solution to get 20 ounces of a 50% solution?

Step 1 Read the problem. Find the amount of the 20% alcohol solution.

Step 2 Assign a variable. Let x = the amount of the 20% alcohol solution.

Step 3 Write an equation.
$$0.20x + 0.12(15) = 0.14(x + 15)$$

Step 4 Solve.
$$0.20x + 1.8 = 0.14x + 2.1$$
$$0.06x = 0.3$$
$$x = 5$$

Step 5 State the answer. 5 gallons of the 20% alcohol solution must be added.

Step 6 Check.
$$0.20(5) + 0.12(15) \overset{?}{=} 0.14(5 + 15)$$
$$2.8 = 2.8$$
The answer checks.

Objective 7 Practice Exercises

For extra help, see Examples 7–8 on pages 72–73 of your text.

Solve each problem.

19. How many pounds of peanuts worth $3 per pound must be mixed with mixed nuts worth $5.50 per pound to make 40 pounds of a mixture worth $5 per pound?

19. _____

20. How many pounds of candy worth $7 per pound must be mixed with candy worth $4.50 per pound to make 100 pounds of candy worth $6 per pound?

20. _____

Chapter 1 LINEAR EQUATIONS, INEQUALITIES, AND APPLICATIONS

1.4 Further Applications of Linear Equations

Learning Objectives
1   Solve problems about different denominations of money.
2   Solve problems about uniform motion.
3   Solve problems about angles.
4   Solve problems about consecutive integers.

Key Terms

Use the vocabulary terms listed below to complete each statement in exercises 1−4.

vertical angles   **complementary angles**   **supplementary angles**

consecutive integers

1. Two angles whose measures sum to 180º are _____.

2. Two angles whose measures sum to 90º are _____.

3. Angles formed by two intersecting lines are called _____.

4. Two integers that differ by 1 are _____.

Objective 1   Solve problems about different denominations of money.

Video Examples

Review this example for Objective 1:

1. A collection of dimes and nickels has a total value of $2.70. The number of nickels is 2 more than twice the number of dimes. How many of each type of coin are in the collection?

    Step 1 Read the problem. Find the number of dimes and the number of nickels.

    Step 2 Assign a variable.
    Let d = the number of dimes.
    Then $2d + 2$ = the number of nickels.

    Step 3 Write an equation.
    $$0.10d + 0.05(2d + 2) = 2.70$$

Now Try:

1. Erica's piggy bank has quarters and nickels in it. The total number of coins in the piggy bank is 50. Their total value is $8.90. How many of each type are in the piggy bank?

    _____

Step 4 Solve.

$$10d + 5(2d + 2) = 270$$
$$10d + 10d + 10 = 270$$
$$20d + 10 = 270$$
$$20d = 260$$
$$d = 13$$

Step 5 State the answer. There are 13 dimes and $2(13) + 2 = 28$ nickels.

Step 6 Check.

$$0.10(13) + 0.05(28) = 1.30 + 1.40 = 2.70$$

The answer is correct.

Objective 1 Practice Exercises

For extra help, see Example 1 on page 79 of your text.

Solve each problem.

1. Anthony sells two different size jars of peanut butter. The large size sells for $2.60 and the small size sells for $1.80. He has 80 jars worth $164. How many of each size jar does he have?

   1.
   large _____
   small _____

2. Stan has 14 bills in his wallet worth $95 altogether. If the wallet contains only $5 and $10 bills, how many bills of each denomination does he have?

   2.
   $5 bills _____
   $10 bills _____

3. Twice as many general admission tickets to a basketball game were sold as reserved seat tickets. General admission tickets cost $10 and reserved seat tickets cost $15. If the total value of both kinds of tickets was $26,250, how many tickets of each kind were sold?

   3.
   general admission_____
   reserved seats _____

Objective 2 Solve problems about uniform motion.

Video Examples

Review these examples for Objective 2:

2. Bill and Hillary start in Washington and fly in opposite directions. At the end of 4 hours, they are 4896 kilometers apart. If Bill flies 60 kilometers per hour faster than Hillary, what are their speeds?

    Step 1 Read the problem. Find the rate of each plane.

    Step 2 Assign a variable.
    Let x = the rate of Hillary's plane.
    Then $x + 60$ = the rate of Bill's plane.
    Using $d = rt$, Hillary's plane goes a distance of $4x$ and Bill's plane goes a distance of $4(x + 60)$.

    Step 3 Write an equation for the total distance.
    $$4x + 4(x + 60) = 4896$$

    Step 4 Solve.
    $$4x + 4x + 240 = 4896$$
    $$8x + 240 = 4896$$
    $$8x = 4656$$
    $$x = 582$$

    Step 5 State the answer. Hillary's plane is flying at 582 kph, and Bill's plane is flying at $582 + 60 = 642$ kph.

    Step 6 Check.
    $$4(582) + 4(642) = 4896$$
    The answer is correct.

3. John left Louisville at noon on the same day that Mike left Louisville at 1 P.M. Both were traveling in the same direction. At 5 P.M., Mike was 62 miles behind John. If John was traveling 2 miles per hour faster than Mike, what were their speeds?

    Step 1 Read the problem. Find the speed for Mike and the speed for John.

    Step 2 Assign a variable. Let x = Mike's speed.
    Then $x + 2$ = John's speed.
    Using $d = rt$, John goes a distance of $5(x + 2)$ and Mike goes a distance of $4x$.

Now Try:

2. Two planes left Philadelphia traveling in opposite directions. Plane A left 15 minutes before plane B. After plane B had been flying for 1 hour, the planes were 860 miles apart. What were the speeds of the two planes if plane A was flying 40 miles per hour faster than plane B?

    _____

3. In a run for charity Ted runs at a speed of 5 miles per hour. Bob leaves 10 minutes after Ted and runs at 6 miles per hour. How long will it take Bob to catch up with Ted? (Hint: Change minutes to hours.)

    _____

Step 3 Write an equation.
$$5(x+2)-4x=62$$

Step 4 Solve.
$$5x+10-4x=62$$
$$x+10=62$$
$$x=52$$

Step 5 State the answer. Mike's speed was 52 mph, and John's speed was 52 + 2 = 54 mph.

Step 6 Check. $5(54)-4(52)=62$ The answer checks.

Objective 2 Practice Exercises

For extra help, see Examples 2–3 on pages 80–81 of your text.

Solve each problem.

4. It takes a kayak $1\frac{1}{2}$ hours to go 24 miles downstream and 4 hours to return. Find the speed of the current and the speed of the kayak in still water.

    4.

    kayak speed _____

    current speed _____

5. A plane can travel 300 miles per hour with the wind and 230 miles per hour against the wind. Find the speed of the wind and the speed of the plane in still air.

    5.

    plane speed _____

    wind speed _____

6. At the beginning of a fund-raising walk, Steve and Vic are 30 miles apart. If they leave at the same time and walk in the same direction, Steve would overtake Vic in 15 hours. If they walked toward each other, they would meet in 3 hours. What are their speeds?

    6.

    Steve _____

    Vic _____

Name: Date:
Instructor: Section:

Objective 3   Solve problems about angles.

Video Examples

Review this example for Objective 3:

4.   Find the measure of each angle.

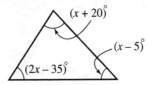

Step 1 Read the problem. Find the measure of each angle.

Step 2 Assign a variable. The variables have been assigned.

Step 3 Write an equation. The sum of the three angles must be 180°.
$$(2x-35)+(x+20)+(x-5)=180$$

Step 4 Solve.
$$2x-35+x+20+x-5=180$$
$$4x-20=180$$
$$4x=200$$
$$x=50$$

Step 5 State the answer. The measures of the angles are $2(50)-35=65°$, $50+20=70°$, and $50-5=45°$.

Step 6 Check. $65°+70°+45°=180°$.

Now Try:

4.   Find the measure of each angle.

_____

Objective 3 Practice Exercises

For extra help, see Example 4 on page 82 of your text.

Solve each problem.

7.   One angle of a triangle measures 30° larger than a second angle. The third angle measures 4 times the second angle. Find the measure of each angle.

7.   _____

8.   The measure of an angle is 15° larger than 10 times its supplement. Find the measure of each angle.

8.   _____

9. The measure of one angle of a triangle is 30° more   9. _____
    than that of a second angle. The measure of the third
    angle is half the measure of the second angle. Find
    the measure of each angle.

Objective 4   Solve problems about consecutive integers.

Video Examples

Review this example for Objective 4:

5. Find three consecutive integers such that the sum
    of the first and third, increased by 2, is 13 more
    than the second.

    Step 1 Read the problem. Find the measure of
    each integer.

    Step 2 Assign a variable. Let
        x = the least of the three consecutive integers.
    Then $x + 1$ = the middle integer,
    and $x + 2$ = the greatest

    Step 3 Write an equation. The sum of the first
    and third, increased by 2, is 13 more than the
    second.
        $$x + (x+2) + 2 = x + 1 + 13$$

    Step 4 Solve.
    $$x + (x+2) + 2 = x + 1 + 13$$
    $$2x + 4 = x + 14$$
    $$x = 10$$

    Step 5 State the answer. The solution is 10, so
    the first integer is $x = 10$, the second is
    $10 + 1 = 11$, and the third is $10 + 2 = 12$. The
    three integers are 10, 11, and 12.

    Step 6 Check. The sum of the first and third is
    $10 + 12 = 22$. If this is increased by 2, the result
    is $22 + 2 = 24$, which is 13 more than the second
    (11). The answer is correct.

Now Try:

5. Find three consecutive integers
    such that the sum of the first and
    second, decreased by 7, is 45
    more than the third.

    _____

Objective 4 Practice Exercises

For extra help, see Example 5 on page 83 of your text.

Solve each problem.

10. Find three consecutive odd integers whose sum is   **10.** _____
     363.

11. The sum of four consecutive even integers is 4. Find   **11.** _____
     the integers.

Chapter 1 LINEAR EQUATIONS, INEQUALITIES, AND APPLICATIONS

1.5   Linear Inequalities in One Variable

Learning Objectives

1   Graph intervals on a number line.
2   Solve linear inequalities using the addition property.
3   Solve linear inequalities using the multiplication property.
4   Solve linear inequalities with three parts.
5   Solve applied problems using linear inequalities.

Key Terms

Use the vocabulary terms listed below to complete each statement in exercises 1−5.

> **interval**   **interval notation**   **inequality**
>
> **linear inequality in one variable**   **equivalent inequalities**

1.   The _____ for $a \leq x < b$ is $[a, b)$.

2.   A(n) _____ can be written in the form $Ax + B > C$, $Ax + B \geq C$, $Ax + B < C$, or $Ax + B \leq C$, where A, B, and C are real numbers with $A \neq 0$.

3.   An algebraic expression related by $>$, $\geq$, $<$, or $\leq$ is called a(n) _____.

4.   An _____ is a portion of a number line.

5.   Inequalities with the same solution set are _____.

Objective 1   Graph intervals on a number line.

Video Examples

Review these examples for Objective 1:

1. Write each inequality in interval notation and graph the interval.

   a. Graph $x > -3$.

   The statement $x > -3$ says that x can represent any value greater than -3, but cannot equal -3, written $(-3, \infty)$. We graph this interval by placing a parenthesis at -3 and drawing an arrow to the right. The parenthesis indicates that -3 is not part of the graph.

Now Try:

1. Write each inequality in interval notation and graph the interval.

   a. Graph $x > -1$.

   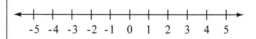

b. Write the inequality $-4 \leq x < 3$ in interval notation, and graph the interval.

$-4 \leq x < 3$

x is between –4 and 3 (excluding 3).

In interval notation, we write $[-4,\ 3)$.

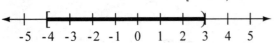

b. Write the inequality $-5 < x \leq -1$ in interval notation, and graph the interval.

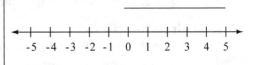

Objective 1 Practice Exercises

For extra help, see Example 1 on pages 90–91 of your text.

Write each inequality in interval notation and graph the interval.

1. $3 < a$

1. _____

2. $y \geq -2$

2. _____

3. $-3 < a \leq 2$

3. _____

Objective 2   Solve linear inequalities using the addition property.

Video Examples

Review this example for Objective 2:

2.   Solve $x - 6 < -11$ and graph the solution set.

$x - 6 < -11$

$x - 6 + 6 < -11 + 6$

$x < -5$

A check confirms that $(-\infty, -5)$, graphed below, is the solution set.

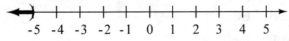

Now Try:

2.   Solve $x + 3 < -1$ and graph the solution set.

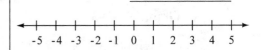

Objective 2 Practice Exercises

For extra help, see Examples 2–3 on pages 92–93 of your text.

Solve each inequality, giving its solution set in both interval and graph forms. Check your answers.

4. $5a + 3 \le 6a$

4. _____

5. $6 + 3x < 4x + 4$

5. _____

6. $3 + 5p \le 4p + 3$

6. _____

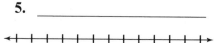

Objective 3   Solve linear inequalities using the multiplication property.

Video Examples

Review these examples for Objective 3:

4.   Solve the inequality, and graph the solution set.

$6x < -24$

We divide each side by 6.
   $6x < -24$

   $\dfrac{6x}{6} < \dfrac{-24}{6}$

   $x < -4$

The graph of the solution set $(-\infty, -4)$, is shown below.

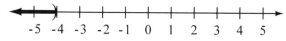

Now Try:

4.   Solve the inequality, and graph the solution set.
$8x \le -40$

5. Solve $-5(x+3)+3 \geq 8-3x$, and graph the solution set.

Step 1 $-5(x+3)+3 \geq 8-3x$

$$-5x-15+3 \geq 8-3x$$

$$-5x-12 \geq 8-3x$$

Step 2 $-5x-12+3x \geq 8-3x+3x$

$$-2x-12 \geq 8$$

$$-2x-12+12 \geq 8+12$$

$$-2x \geq 20$$

Step 3 $\dfrac{-2x}{-2} \leq \dfrac{20}{-2}$

$$x \leq -10$$

The solution set is $(-\infty, -10]$. The graph is shown below.

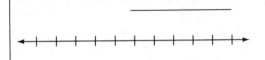

5. Solve $-6(x+4)+1 \geq 9-2x$, and graph the solution set.

_____

6. Solve $-\dfrac{2}{5}(r-4)-\dfrac{1}{3} < \dfrac{1}{3}(4-r)$, and graph the solution set.

Step 1 $15\left[-\dfrac{2}{5}(r-4)-\dfrac{1}{3}\right] < 15\left[\dfrac{1}{3}(4-r)\right]$

$$15\left[-\dfrac{2}{5}(r-4)\right]-15\left(\dfrac{1}{3}\right) < 15\left[\dfrac{1}{3}(4-r)\right]$$

$$-6(r-4)-5 < 5(4-r)$$

$$-6r+24-5 < 20-5r$$

$$-6r+19 < 20-5r$$

Step 2 $-6r+19+5r < 20-5r+5r$

$$-r+19 < 20$$

$$-r+19-19 < 20-19$$

$$-r < 1$$

Step 3 $-1(-r) > -1(1)$

$$r > -1$$

The solution set is $(-1, \infty)$. The graph is shown below.

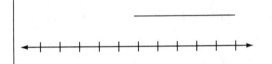

6. Solve $-\dfrac{1}{2}(r-5)+\dfrac{1}{3} < \dfrac{1}{3}(5-r)$, and graph the solution set.

_____

Objective 3 Practice Exercises

For extra help, see Examples 4–6 on pages 94–96 of your text.

Solve each inequality, giving its solution set in both interval and graph forms. Check your answers.

7. $-2s < 4$

7. _____

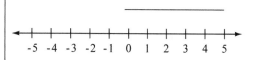

8. $4k \geq -16$

8. _____

9. $-9m \geq -36$

9. _____

Objective 4   Solve linear inequalities with three parts.

Video Examples

Review this example for Objective 4:

8.   Solve $3 \leq 4x - 5 < 7$ and graph the solution set.

$$3 \leq -4x - 5 < 7$$
$$3 + 5 \leq -4x < 7 + 5$$
$$8 \leq -4x < 12$$
$$-\frac{8}{4} \geq -\frac{4x}{4} > -\frac{12}{4}$$
$$-2 \geq x > -3$$
$$-3 < x \leq -2$$

The solution set is $(-3, -2]$. The graph is shown below.

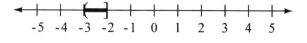

Now Try:

8.   Solve $8 \leq -6x - 4 < 20$ and graph the solution set.

_____

Objective 4 Practice Exercises

For extra help, see Examples 7–8 on page 97 of your text.

Solve each inequality, giving its solution set in both interval and graph forms. Check your answers.

10. $7 < 2x + 3 \leq 13$

10. _____

11. $-17 \leq 3x - 2 < -11$

11. _____

12. $1 < 3z + 4 < 19$

12. _____

Objective 5   Solve applied problems using linear inequalities.

Video Examples

Review this example for Objective 5:

10. Ruth tutors mathematics in the evenings in an office for which she pays $600 per month rent. If rent is her only expense and she charges each student $40 per month, how many students must she teach to make a profit of at least $1600 per month?

Step 1 Read the problem again.

Step 2 Assign a variable.
   Let $x =$ the number of students.

Step 3 Write an inequality.
   $40x - 600 \geq 1600$

Step 4 Solve.
   $40x - 600 + 600 \geq 1600 + 600$

   $\qquad\qquad 40x \geq 2200$

   $\qquad\qquad \dfrac{40x}{40} \geq \dfrac{2200}{40}$

   $\qquad\qquad\quad x \geq 55$

Now Try:

10. Two sides of a triangle are equal in length, with the third side 8 feet longer than one of the equal sides. The perimeter of the triangle cannot be more than 38 feet. Find the largest possible value for the length of the equal sides.

_____

Step 5 State the answer. Ruth must have 55 or more students to have at least $1600 profit.

Step 6 Check. $40(55) - 600 = 1600$ Also, any number greater than 55 makes the profit greater than $1600.

Objective 5 Practice Exercises

For extra help, see Examples 9–10 on pages 98–99 of your text.

Solve each problem.

13. Lauren has grades of 98 and 86 on her first two chemistry quizzes. What must she score on her third quiz to have an average of at least 91 on the three quizzes?

13. _____

14. Nina has a budget of $230 for gifts for this year. So far she has bought gifts costing $47.52, $38.98, and $26.98. If she has three more gifts to buy, find the average amount she can spend on each gift and still stay within her budget.

14. _____

15. Andrew must have an average of 80 of the points on four exams to receive a B in the class. He has earned 78, 83, and 75 on the first three exams. What is the lowest score he can earn on a 100-point test to guarantee a B in the class?

15. _____

Chapter 1 LINEAR EQUATIONS, INEQUALITIES, AND APPLICATIONS

1.6   Set Operations and Compound Inequalities

Learning Objectives

1   Recognize set intersection and union.
2   Find the intersection of two sets.
3   Solve compound inequalities with the word *and*.
4   Find the union of two sets.
5   Solve compound inequalities with the word *or*.

Key Terms

Use the vocabulary terms listed below to complete each statement in exercises 1−3.

     intersection   **compound inequality**   **union**

1. The _____ of two sets, A and B, is the set of elements that belong to either A or B or both.

2. A _____ is formed by joining two inequalities with a connective word such as *and* or *or*.

3. The _____ of two sets, A and B, is the set of elements that belong to both A and B.

Objective 2   Find the intersection of two sets.

Video Examples

Review this example for Objective 2:

1. Let $A = \{6, 7, 8, 9\}$ and $B = \{6, 8, 10\}$. Find $A \cap B$.

   The set $A \cap B$ contains those elements that belong to both A and B.
   $$A \cap B = \{6, 7, 8, 9\} \cap \{6, 8, 10\}$$
   $$= \{6, 8\}$$

Now Try:

1. Let $A = \{20, 30, 40, 50\}$ and $B = \{30, 50, 70\}$. Find $A \cap B$.

   _____

Objective 2 Practice Exercises

For extra help, see Example 1 on page 103 of your text.

Let $A = \{0, 1, 2, 3, 4, 5\}$, $B = \{2, 4, 6, 8, 10\}$, $C = \{1, 3, 5, 7, 9\}$, and $D = \{0, 2, 4\}$. Specify each set.

1. $A \cap D$ 1. _____

2. $B \cap C$ 2. _____

3. $A \cap C$ 3. _____

Objective 3   Solve compound inequalities with the word *and*.

Video Examples

Review these examples for Objective 3:

2. Solve $x+6 \leq 15$ and $x+3 \geq 7$, and graph the solution set.

Step 1 Solve each inequality individually.

$\quad x+6 \leq 15 \qquad$ and $\qquad x+3 \geq 7$

$x+6-6 \leq 15-6 \quad$ and $\quad x+3-3 \geq 7-3$

$\qquad x \leq 9 \qquad$ and $\qquad x \geq 4$

Step 2 The solution set of the compound inequality includes all numbers that satisfy both inequalities from Step 1.

The graphs of $x \leq 9$ and $x \geq 4$ are shown below.

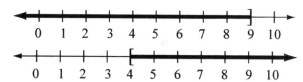

The intersection of these two graphs is the solution set [4, 9].

3. Solve $-5x-6 > 8$ and $6x-3 \leq -21$, and graph the solution set.

Step 1 Solve each inequality individually.

$-5x-6 > 8 \qquad$ and $\quad 6x-3 \leq -21$

$\quad -5x > 14 \qquad$ and $\qquad 6x \leq -18$

$\qquad x < -\dfrac{14}{5} \quad$ and $\qquad x \leq -3$

The graphs of $x < -\dfrac{14}{5}$ and $x \leq -3$ are shown

below.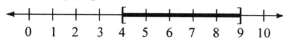

Now Try:

2. Solve $x+4 \leq 20$ and $x-3 \geq 9$ and graph the solution set.

_____

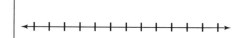

3. Solve $-8x+6 > 30$ and $4x-7 \leq 11$, and graph the solution set.

_____

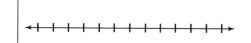

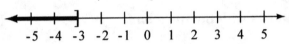

Step 2 Now find all the values of x that are less than $-\dfrac{14}{5}$ and also less than or equal to -3. The solution set is $(-\infty, -3]$.

Objective 3 Practice Exercises

For extra help, see Examples 2–4 on pages 104–105 of your text.

For each compound inequality, give the solution set in both interval and graph forms.

4. $1 - 2s \le 7$ and $2s + 7 \ge 11$

4. _____

5. $3x + 2 < 11$ and $2 - 3x \le 14$

5. _____

6. $5t > 0$ and $5t + 4 \le 9$

6. _____

Objective 4   Find the union of two sets.

Video Examples

Review this example for Objective 4:

5. Let $A = \{6, 7, 8, 9\}$ and $B = \{6, 8, 10\}$. Find $A \cup B$.

The set $A \cup B$ contains those elements that belong to either A or B (or both).

$A \cup B = \{6, 7, 8, 9\} \cup \{6, 8, 10\}$

$\qquad\quad = \{6, 7, 8, 9, 10\}$

Now Try:

5. Let $A = \{20, 30, 40, 50\}$ and $B = \{30, 50, 70\}$. Find $A \cup B$.

    _____

Objective 4 Practice Exercises

For extra help, see Example 5 on page 105 of your text.

Let A = {0, 1, 2, 3, 4, 5}, *B* = {2, 4, 6, 8, 10}, *C* = {1, 3, 5, 7, 9}, *D* = {0, 2, 4}, *and*
E = {0}. *Specify each set.*

7. $A \cup D$ 7. _____

8. $B \cup C$ 8. _____

9. $A \cup E$ 9. _____

Objective 5   Solve compound inequalities with the word *or*.

Video Examples

Review these examples for Objective 5:

6. Solve the compound inequality, and graph the solution set.

$$7x - 6 < 4x \quad \text{or} \quad -4x \leq -16$$

Step 1 Solve each inequality individually.
$$7x - 6 < 4x \quad \text{or} \quad -4x \leq -16$$
$$3x < 6$$
$$x < 2 \quad \text{or} \quad x \geq 4$$
The graphs of $x < 2$ and $x \geq 4$ are shown below.

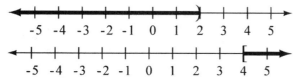

Step 2 Since the inequalities are joined with *or*, find the union of the two solution sets. The union is shown below. $(-\infty, 2) \cup [4, \infty)$

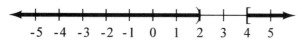

Now Try:

6. Solve the compound inequality, and graph the solution set.
$$5x - 6 < 2x \quad \text{or} \quad -7x \leq -35$$

_____

Name: Date:

Instructor: Section:

8. Solve the compound inequality, and graph the solution set.

$$-5x + 4 \geq -16 \text{ or } 6x - 11 \geq -29$$

Solve each inequality.

$$-5x + 4 \geq -16 \quad \text{or} \quad 6x - 11 \geq -29$$
$$-5x \geq -20 \quad \text{or} \quad \quad 6x \geq -18$$
$$x \leq 4 \quad \quad \text{or} \quad \quad \quad x \geq -3$$

Graph each inequality.

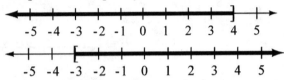

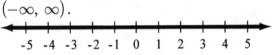

The solution is the union of the two sets,

$(-\infty, \infty)$.

8. Solve the compound inequality, and graph the solution set.

$$-9x + 3 \geq -15 \text{ or } 8x - 7 \geq -15$$

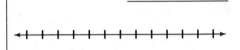

_____

Objective 5 Practice Exercises

For extra help, see Examples 6–9 on pages 106–108 of your text.

For each compound inequality, give the solution set in both interval and graph forms.

10. $q + 3 > 7$ or $q + 1 \leq -3$

10. _____

11. $3 > 4m + 2$ or $4m - 3 \geq -2$

11. _____

12. $2r + 4 \geq 8$ or $4r - 3 < 1$

12. _____

Chapter 1 LINEAR EQUATIONS, INEQUALITIES, AND APPLICATIONS

1.7   Absolute Value Equations and Inequalities

Learning Objectives	
1	Use the distance definition of absolute value.
2	Solve equations of the form $|ax + b| = k$, for $k > 0$.
3	Solve inequalities of the form $|ax + b| < k$ and of the form $|ax + b| > k$, for $k > 0$.
4	Solve absolute value equations that involve rewriting.
5	Solve equations of the form $|ax + b| = |cx + d|$.
6	Solve special cases of absolute value equations and inequalities.
7	Solve an application involving relative error.

Key Terms

Use the vocabulary terms listed below to complete each statement in exercises 1−2.

absolute value equation   **absolute value inequality**

1.   An _____ is an equation that involves the absolute value of a variable expression.

2.   An _____ is an inequality that involves the absolute value of a variable expression.

Objective 2   Solve equations of the form $|ax + b| = k$, for $k > 0$.

Video Examples

Review this example for Objective 2:

1.   Solve $|3x + 2| = 14$. Graph the solution set.

This is Case 1.

$$3x + 2 = 14 \quad \text{or} \quad 3x + 2 = -14$$
$$3x = 12 \quad \text{or} \quad 3x = -16$$
$$x = 4 \quad \text{or} \quad x = -\frac{16}{3}$$

Check

Let $x = 4$ Let $x = -\frac{16}{3}$

$$3(4) + 2 \overset{?}{=} 14 \quad \Big| \quad 3\left(-\frac{16}{3}\right) + 2 \overset{?}{=} -14$$
$$12 + 2 \overset{?}{=} 14 \quad \Big| \quad -16 + 2 \overset{?}{=} -14$$
$$14 = 14 \quad \Big| \quad -14 = -14$$

Now Try:

1.   Solve $|5x + 4| = 11$. Graph the solution set.

_____

The check confirms that the solution set is

$\left\{-\dfrac{16}{3}, 4\right\}$.

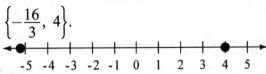

Objective 2 Practice Exercises

For extra help, see Example 1 on page 113 of your text.

Solve each equation.

1. $|2x+3|=10$

                                                         1. _____

2. $|5r-15|=0$

                                                         2. _____

3. $\left|\dfrac{1}{2}x-3\right|=4$

                                                         3. _____

Objective 3   **Solve inequalities of the form $|ax + b| < k$ and of the form $|ax + b| > k$, for $k > 0$.**

Video Examples

Review these examples for Objective 3:

2.   Solve $|4x+2|>10$. Graph the solution set.

     This is Case 2.

$$4x+2>10 \quad \text{or} \quad 4x+2<-10$$
$$4x>8 \quad\;\; \text{or} \quad\quad 4x<-12$$
$$x>2 \quad\;\; \text{or} \quad\quad\;\; x<-3$$

     A check confirms that the solution set is

     $(-\infty,-3)\cup(2,\infty)$.

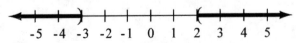

Now Try:

2.   Solve $|8x+4|>12$. Graph the solution set.

_____

3. Solve $|10x+5|<25$. Graph the solution set.

This is Case 3.

$$-25<10x+5<25$$

$$-30<\ \ 10x\ \ <20$$

$$-3<\ \ \ x\ \ \ <2$$

A check confirms that the solution set is $(-3, 2)$.

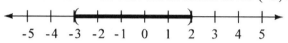

3. Solve $|6x+3|<15$. Graph the solution set.

Objective 3 Practice Exercises

For extra help, see Examples 2–4 on pages 113–115 of your text.

Solve each inequality and graph the solution set.

4. $|x-2|>8$

4. _____

5. $|2r-9|\geq 23$

5. _____

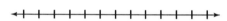

6. $|5r+2|<18$

6. _____

Objective 4   Solve absolute value equations that involve rewriting.

Video Examples

Review this example for Objective 4:

5. Solve $|x-7|+3=15$.

Isolate the absolute value.

$$|x-7|+3=15$$

$$|x-7|+3-3=15-3$$

$$|x-7|=12$$

Now Try:

5. Solve $|x-3|+4=11$.

_____

Now solve, using Case 1.

$$x - 7 = 12 \quad \text{or} \quad x - 7 = -12$$
$$x = 19 \quad \text{or} \quad x = -5$$

Check

Let $x = 19$ Let $x = -5$

$$|19 - 7| + 3 \overset{?}{=} 15 \quad | \quad |-5 - 7| + 3 \overset{?}{=} 15$$

$$|12| + 3 \overset{?}{=} 15 \quad | \quad |-12| + 3 \overset{?}{=} 15$$

$$15 = 15 \quad | \quad 15 = 15$$

The check confirms that the solution set is $\{-5, 19\}$.

Objective 4 Practice Exercises

For extra help, see Examples 5–6 on page 115 of your text.

Solve each equation.

7. $|2w - 1| + 7 = 12$ 7. _____

8. $\left| 2 - \dfrac{1}{2}x \right| - 5 = 18$ 8. _____

9. $|4t + 3| + 8 = 10$ 9. _____

Objective 5   Solve equations of the form $|ax + b| = |cx + d|$.

Video Examples

Review this example for Objective 5:

7.   Solve $|z+5| = |3z-9|$.

$$z+5 = 3z-9 \quad \text{or} \quad z+5 = -(3z-9)$$
$$z+14 = 3z \quad\quad \text{or} \quad z+5 = -3z+9$$
$$14 = 2z \quad\quad\quad \text{or} \quad\quad 4z = 4$$
$$z = 7 \quad\quad\quad\quad \text{or} \quad\quad\quad z = 1$$

Check

Let $z = 7$ Let $z = 1$

$$|7+5| \overset{?}{=} |3(7)-9| \quad\quad |1+5| \overset{?}{=} |3(1)-9|$$
$$|12| \overset{?}{=} |21-9| \quad\quad\quad |6| \overset{?}{=} |-6|$$
$$12 = 12 \quad\quad\quad\quad\quad 6 = 6$$

The check confirms that the solution set is $\{1, 7\}$.

Now Try:

7.   Solve $|z+7| = |2z+8|$.

_____

Objective 5 Practice Exercises

For extra help, see Example 7 on page 116 of your text.

Solve each problem.

10. $|y+5| = |3y+1|$ **10.** _____

11. $|2p-4| = |7-p|$ **11.** _____

12. $|3x-2| = |5x+8|$ **12.** _____

Objective 6   Solve special cases of absolute value equations and inequalities.

Video Examples

Review these examples for Objective 6:

8.   Solve each equation.

a. $|9x - 1| = -13$

The absolute value of an expression can never be negative, so there are no solutions for this equation. The solution set is $\varnothing$.

b. $|8x - 12| = 0$

$$8x - 12 = 0$$
$$8x = 12$$
$$x = \frac{3}{2}$$

The solution set is $\left\{ \frac{3}{2} \right\}$.

9.   Solve each inequality.

a. $|x| \geq -12$

The absolute value of a number is always greater than or equal to 0. Thus, $|x| \geq -12$ is always true, and the solution set is $(-\infty, \infty)$.

b. $|x - 11| + 7 < 2$

$$|x - 11| + 7 < 2$$
$$|x - 11| < -5$$

There is no number whose absolute value is less than -5, so this inequality has no solution. The solution set is $\varnothing$.

c. $|2x - 8| - 5 \leq -5$

$$|2x - 8| - 5 \leq -5$$
$$|2x - 8| \leq 0$$

The value of $|2x - 8|$ will never be less than zero. However, $|2x - 8|$ will equal 0.

$$2x - 8 = 0$$
$$2x = 8$$
$$x = 4$$

The solution set is $\{4\}$.

Now Try:

8.   Solve each equation.

a. $|6x + 13| = -5$

_____

b. $|7x + 35| = 0$

_____

9.   Solve each inequality.

a. $|x| \geq -2$

_____

b. $|x + 12| + 6 < 3$

_____

c. $|5x - 20| + 7 \leq 7$

_____

Objective 6 Practice Exercises

For extra help, see Examples 8–9 on pages 116–117 of your text.

Solve each problem.

13. $\left|7+\dfrac{1}{2}x\right|=0$

13. _____

14. $\left|m-2\right|\geq -1$

14. _____

15. $\left|k+5\right|\leq -2$

15. _____

Objective 7   Solve an application involving relative error.

Video Examples

Review this example for Objective 7:

10. Suppose a machine filling 16.9 oz water bottles is set for a relative error that is no greater than 0.025 oz. How many ounces may a filled water bottle contain?

$$\left|\frac{16.9-x}{16.9}\right|\leq 0.025$$

$$-0.025\leq \frac{16.9-x}{16.9}\leq 0.025$$

$$-0.4225\leq 16.9-x\leq 0.4225$$

$$-17.3225\leq \ -x\ \leq -16.4775$$

$$17.3225\geq \quad x\ \geq 16.4775$$

$$16.4775\leq \quad x\ \leq 17.3225$$

The bottle may contain between 16.4775 and 17.3225 oz, inclusive.

Now Try:

10. Suppose a machine filling 16.9 oz water bottles is set for a relative error that is no greater than 0.05 oz. How many ounces may a filled water bottle contain?

_____

For extra help, see Example 10 on pages 117–118 of your text.

Determine the number of ounces a filled 16.9 oz water bottle may contain for the given relative error.

16.   no greater than 0.04 oz

16. _____

17.   no greater than 0.015 oz

17. _____

18.   no greater than 0.03 oz

17. _____

Chapter 2 LINEAR EQUATIONS, GRAPHS, AND FUNCTIONS

2.1   Linear Equations in Two Variables

Learning Objectives	
1	Interpret a line graph.
2	Plot ordered pairs.
3	Find ordered pairs that satisfy a given equation.
4	Graph lines.
5	Find x- and y-intercepts.
6	Recognize equations of horizontal and vertical lines.
7	Use the midpoint formula.

Key Terms

Use the vocabulary terms listed below to complete each statement in exercises 1−14.

ordered pair	origin	x-axis	y-axis
rectangular (Cartesian) coordinate system			plot
components	coordinate	quadrant	graph of an equation
first-degree equation		linear equation in two variables	
y-intercept	x-intercept		

1.   If a graph intersects the y-axis at k, then the _____ is $(0, k)$.

2.   An equation that can be written in the form $Ax + By = C$, where A, B, and C are real numbers and $A, B \neq 0$, is called a _____.

3.   Each number in an ordered pair represents a _____ of the corresponding point.

4.   The axis lines in a coordinate system intersect at the _____.

5.   If a graph intersects the x-axis at k, then the _____ is $(k, 0)$.

6.   In a rectangular coordinate system, the horizontal number line is called the
     _____.

7.   A _____ is one of the four regions in the plane determined by a rectangular coordinate system.

8.   A pair of numbers written between parentheses in which order is important is called a(n) _____.

9.   In a rectangular coordinate system, the vertical number line is called the
     _____.

10.   The two numbers in an ordered pair are the _____ of the ordered pair.

11.   To _____ an ordered pair is to locate the corresponding point on a coordinate system.

12.   Together, the x-axis and the y-axis form a _____.

13.   The _____ is the set of points corresponding to all ordered pairs that satisfy the equation.

14.   A _____ has no term with a variable to a power greater than one.

Objective 3   Find ordered pairs that satisfy a given equation.

Video Examples

Review these examples for Objective 3:

1.   In parts (a)–(d), complete each ordered pair for $5x + 4y = 20$. Then part (e), write the results as a table of ordered pairs.

   a. $(0, \underline{})$

   $5x + 4y = 20$

   $5(0) + 4y = 20$

   $4y = 20$

   $y = 5$

   The ordered pair is $(0, 5)$.

   b. $(\underline{}, 0)$

   $5x + 4y = 20$

   $5x + 4(0) = 20$

   $5x = 20$

   $x = 4$

   The ordered pair is $(4, 0)$.

   c. $(-4, \underline{})$

   $5x + 4y = 20$

   $5(-4) + 4y = 20$

   $-20 + 4y = 20$

   $4y = 40$

   $y = 10$

   The ordered pair is $(-4, 10)$.

Now Try:

1.   In parts (a)–(d), complete each ordered pair for $2x - 5y = 10$. Then part (e), write the results as a table of ordered pairs.

   a. $(0, \underline{})$

   _____

   b. $(\underline{}, 0)$

   _____

   c. $(-5, \underline{})$

   _____

d. (___,–5)

$$5x + 4y = 20$$

$$5x + 4(-5) = 20$$

$$5x - 20 = 20$$

$$5x = 40$$

$$x = 8$$

The ordered pair is (8,–5).

e. Write the ordered pairs in a table.

x	y
0	5
4	0
−4	10
8	−5

d. (___, 2)

_____

e. Write the ordered pairs in a table.

_____

Objective 3 Practice Exercises

For extra help, see Example 1 on page 138 of your text.

For each of the given equations, complete the ordered pairs beneath it.

1. $5x + 4y = 10$

    (a)   (2,)

    (b)   (4,)

    (c)   (,3)

    (d)   (0,)

    (e)   (,2)

1.

    (a) _____

    (b) _____

    (c) _____

    (d) _____

    (e) _____

Complete each table of values. Write the results as ordered pairs.

2. $-7x + 2y = -14$

x	y
	0
0	
3	
	7

2. _____

3. $y - 4 = 0$ **3.** _____

x	y
-4	
0	
6	
-12	

Objective 5   Find *x*- and *y*-intercepts.

Video Examples

Review these examples for Objective 5:

2. Find the *x*- and *y*-intercepts of $3x + y = 6$ and graph the equation.

To find the *y*-intercept, let $x = 0$.
To find the *x*-intercept, let $y = 0$.

$$3(0) + y = 6 \quad | \quad 3x + 0 = 6$$
$$0 + y = 6 \quad | \quad 3x = 6$$
$$y = 6 \quad | \quad x = 2$$

The intercepts are (0, 6) and (2, 0). To find a third point, as a check, we let $x = 1$.

$$3(1) + y = 6$$
$$3 + y = 6$$
$$y = 3$$

This gives the ordered pair (1, 3).

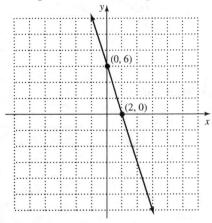

Now Try:

2. Find the *x*- and *y*-intercepts of $5x - 2y = -10$ and graph the equation.

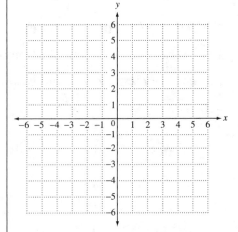

Name: Date:
Instructor: Section:

3. Graph $x + 5y = 0$.

To find the y-intercept, let $x = 0$.
To find the x-intercept, let $y = 0$.

$$0 + 5y = 0 \quad \bigg| \quad x + 5(0) = 0$$
$$5y = 0 \quad \bigg| \quad x + 0 = 0$$
$$y = 0 \quad \bigg| \quad x = 0$$

The x- and y-intercepts are the same point (0, 0).
We must select two other values for x or y to find
two other points. We choose $y = 1$ and $y = -1$.

$$x + 5(1) = 0 \quad \bigg| \quad x + 5(-1) = 0$$
$$x + 5 = 0 \quad \bigg| \quad x - 5 = 0$$
$$x = -5 \quad \bigg| \quad x = 5$$

We use (–5, 1), (0, 0), and (5, –1) to draw the
graph.

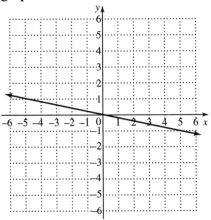

3. Graph $3x - y = 0$.

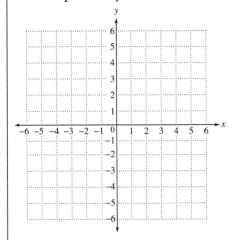

Objective 5 Practice Exercises

For extra help, see Examples 2–3 on pages 140–141 of your text.

Find the intercepts, then graph the equation.

4. $4x - y = 4$

4. _____

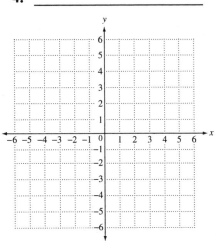

5. $2x - 3y = 6$ 5. _____

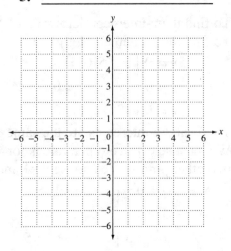

Objective 6   Recognize equations of horizontal and vertical lines.

Video Examples

Review these examples for Objective 6:

4.   Graph each equation.

  a. $y = -2$

  For any value of x, y is always -2. Three
  ordered pairs that satisfy the equation are
  $(-4, -2)$, $(0, -2)$ and $(2, -2)$. Drawing a line
  through these points gives the horizontal line.
  The y-intercept is $(0, -2)$. There is no x-intercept.

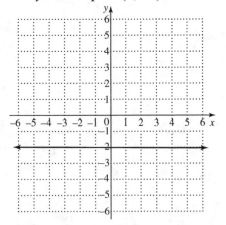

  b. $x + 4 = 0$

  First we subtract 4 from each side of the
  equation to get the equivalent equation $x = -4$.
  All ordered-pair solutions of this equation have
  x-coordinate -4.
  Three ordered pairs that satisfy the equation are
  $(-4, -1)$, $(-4, 0)$, and $(-4, 3)$. The graph is a

Now Try:

4.   Graph each equation.

  a. $y = 4$

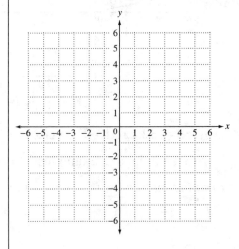

  b. $x = 0$

vertical line. The *x*-intercept is (–4, 0). There is no *y*-intercept.

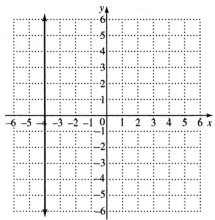

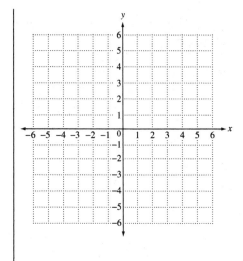

Objective 6 Practice Exercises

For extra help, see Example 4 on pages 141–142 of your text.

Find the intercepts, and graph the line.

6. $x - 1 = 0$

6.

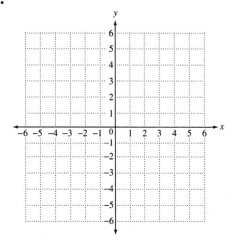

7. $y + 3 = 0$

7.

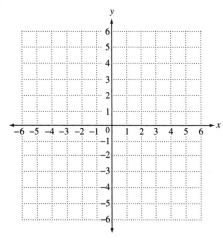

Objective 7   Use the midpoint formula.

Video Examples

Review this example for Objective 7:

5. Find the coordinates of the midpoint of the line segment PQ with endpoints $P(8,-5)$ and $Q(4,-3)$.

$$P(8,-5)=(x_1,\,y_1) \text{ and } Q(4,-3)=(x_2,\,y_2)$$

$$\left(\frac{x_1+x_2}{2},\frac{y_1+y_2}{2}\right)=\left(\frac{8+4}{2},\frac{-5+(-3)}{2}\right)$$

$$=\left(\frac{12}{2},\frac{-8}{2}\right)$$

$$=(6,-4)$$

The midpoint of PQ is $(6,-4)$.

Now Try:

5. Find the coordinates of the midpoint of the line segment PQ with endpoints $P(7,-6)$ and $Q(3, 2)$.

_____

Objective 7 Practice Exercises

For extra help, see Example 5 on page 143 of your text.

Find the midpoint of each segment with the given endpoints.

8.   (–4, 8) and (8,–4)   8. _____

9.   (8, 5) and (–3,–11)   9. _____

10.   (–2.2,–9.3) and (–8.4, 5.7)   10. _____

Chapter 2 LINEAR EQUATIONS, GRAPHS, AND FUNCTIONS

2.2   The Slope of a Line

Learning Objectives
1   Find the slope of a line, given two points on the line.
2   Find the slope of a line, given an equation of the line.
3   Graph a line, given its slope and a point on the line.
4   Use slopes to determine whether two lines are parallel, perpendicular, or neither.
5   Solve problems involving average rate of change.

Key Terms

Use the vocabulary terms listed below to complete each statement in exercises 1−3.

rise   run   slope

1.   The _____ of a line is the ratio of the change in y compared to the change in x when moving along the line from one point to another.

2.   The vertical change between two different points on a line is called the

_____.

3.   The horizontal change between two different points on a line is called the

_____.

Objective 1   Find the slope of a line, given two points on the line.

Video Examples

Review this example for Objective 1:

1.   Find the slope of the line passing through $(-5, 4)$ and $(2, -6)$

Apply the slope formula.
$$(x_1, y_1) = (-5, 4) \text{ and } (x_2, y_2) = (2, -6)$$

$$\text{slope } m = \frac{y_2 - y_1}{x_2 - x_1} = \frac{-6 - 4}{2 - (-5)}$$

$$= \frac{-10}{7}, \text{or } -\frac{10}{7}$$

Now Try:

1.   Find the slope of the line passing through $(-6, 7)$ and $(3, -9)$

_____

Objective 1 Practice Exercises

For extra help, see Example 1 on page 149 of your text.

Find the slope of the line through the given points.

1. $(4, 3)$ and $(3, 5)$

1.   _____

2. (5,–2) and (2, 7)

2. _____

3. (7, 2) and (–7, 3)

3. _____

Objective 2   Find the slope of a line, given an equation of the line.

Video Examples

Review these examples for Objective 2:

2. Find the slope of the line $5x + y = 10$.

The intercepts can be used as the two points needed to find the slope. Let $y = 0$ to find that the x-intercept is (2, 0). Then let $x = 0$ to find that the y-intercept is (0, 10).

$$m = \frac{y_2 - y_1}{x_2 - x_1} = \frac{10 - 0}{0 - 2} = \frac{10}{-2} = -5$$

3. Find the slope of each line.

a. $x + 4 = 0$

The graph of $x + 4 = 0$ is a vertical line.

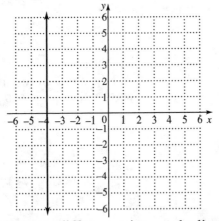

Select two different points on the line, such as (–4, 3) and (–4,–2) and use the slope formula.

$$m = \frac{y_2 - y_1}{x_2 - x_1} = \frac{-2 - 3}{-4 - (-4)} = \frac{-5}{0} \text{ undefined}$$

The slope is undefined.

Now Try:

2. Find the slope of the line $7x - y = 6$.

_____

3. Find the slope of each line.

a. $x = 0$

_____

b. $y = -2$

The graph of $y = -2$ is a horizontal line.

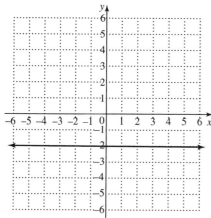

Select two different points on the line, such as $(-4,-2)$ and $(2,-2)$ and use the slope formula.
$$m = \frac{y_2 - y_1}{x_2 - x_1} = \frac{-2-(-2)}{2-(-4)} = \frac{0}{6} = 0$$
The slope is 0.

4. Find the slope of the graph of $4x - 3y = 7$.

Solve the equation for y.
$$4x - 3y = 7$$
$$-3y = -4x + 7$$
$$y = \frac{4}{3}x - \frac{7}{3}$$

The slope is given by the coefficient of x, so the slope is $\frac{4}{3}$.

b. $y = 4$

_____

4. Find the slope of the graph of $7x - 4y = 8$.

_____

Objective 2 Practice Exercises

For extra help, see Examples 2–4 on pages 150–151 of your text.

Find the slope of each line.

4. $x = 0$

4. _____

5. $3y = 2x - 1$

5. _____

6. $2x + 7y = 7$ **6.** _____

Objective 3   Graph a line, given its slope and a point on the line.

Video Examples

Review this example for Objective 3:

5.   Graph the line passing through the point $(1, -3)$, with slope $-\dfrac{5}{2}$.

First, locate the point $(1, -3)$. Then write the slope $-\dfrac{5}{2}$ as

$$\text{slope } m = \frac{\text{change in } y \text{ (rise)}}{\text{change in } x \text{ (run)}} = \frac{5}{-2}.$$

Locate another point on the line by counting up 5 units from $(1, -3)$, and then to the left 2 units. Finally, draw the line through this new point, $(-1, 2)$.

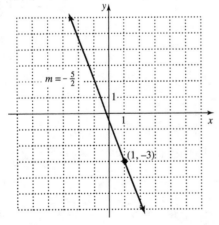

Now Try:

5.   Graph the line passing through the point $(2, 2)$, with slope $\dfrac{1}{3}$.

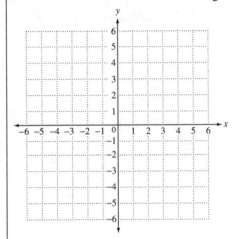

Objective 3 Practice Exercises

For extra help, see Example 5 on page 152 of your text.

Graph the line passing through the given point and having the given slope.

7. $(4, -2)$; $m = -1$

7.

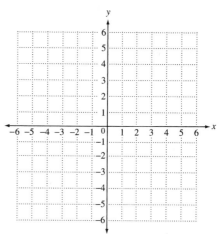

8. $(-3, -2)$; $m = \frac{2}{3}$

8.

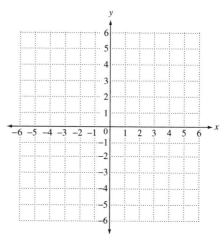

9. $(-3, -1)$; undefined slope

9.

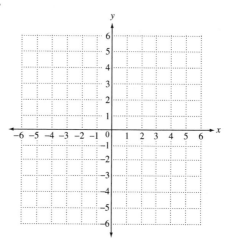

83

Objective 4   Use slopes to determine whether two lines are parallel, perpendicular, or neither.

Video Examples

Review these examples for Objective 4:

6. Determine whether the lines L_1 passing through (7, 6) and (–4, 2) and L_2 passing through (0,–4) and (11, 0), are parallel.

Slope of L_1 : $m_1 = \dfrac{2-6}{-4-7} = \dfrac{-4}{-11} = \dfrac{4}{11}$

Slope of L_2 : $m_2 = \dfrac{0-(-4)}{11-0} = \dfrac{4}{11}$

The slopes are equal, the two lines are parallel.

7. Are the lines with equations $x+3y=8$ and $-3x+y=5$ perpendicular?

Find the slope of each line by first solving each equation for y.

$$3y = -x+8 \qquad \bigg| \qquad y = 3x+5$$

$$y = -\dfrac{1}{3}x+\dfrac{8}{3}$$

The slope is $-\dfrac{1}{3}$. $\bigg|$ The slope is 3.

Check the product of the slopes: $-\dfrac{1}{3}(3) = -1$.

The two lines are perpendicular because the product of their slopes is –1.

Now Try:

6. Determine whether the lines L_1 passing through (6, 3) and (–4, 5) and L_2 passing through (0, 3) and (15, 0), are parallel.

_____

7. Are the lines with equations $9x-y=7$ and $x+9y=11$ perpendicular?

_____

Objective 4 Practice Exercises

For extra help, see Examples 6–8 on pages 153–154 of your text.

Decide whether the lines in each pair are **parallel**, **perpendicular**, *or* **neither**.

10. $y = -5x-2$
        $y = 5x+11$

10. _____

11. $-x+y=-7$
        $x-y=-3$

11. _____

12. $2x + 2y = 7$

      $2x - 2y = 5$

12. _____

Objective 5   Solve problems involving average rate of change.

Video Examples

Review these examples for Objective 5:

9. A small company had the following sales during their first three years of operation.

Year	Sales
2005	$82,250
2006	$89,790
2007	$96,100

a. What was the rate of change in 2005–2006?

b. What was the rate of change in 2006–2007?

c. What was the rate of change in 2005–2007?

a. We use the ordered pairs (2005, 82,250) and (2006, 89,790).

average rate of change $= \dfrac{89,790 - 82,250}{2006 - 2005}$

$= \dfrac{7540}{1} = 7540$

This means sales increased by an average of $7540 from 2005 to 2006.

b. We use the ordered pairs (2006, 89,790) and (2007, 96,100).

average rate of change $= \dfrac{96,100 - 89,790}{2007 - 2006}$

$= \dfrac{6310}{1} = 6310$

This means sales increased by an average of $6310 from 2006 to 2007.

c. We use the ordered pairs (2005, 82,250) and (2007, 96,100).

average rate of change $= \dfrac{96,100 - 82,250}{2007 - 2005}$

$= \dfrac{13,850}{2} = 6925$

This means sales increased by an average of $6925 from 2005 to 2007.

Now Try:

9. A plane had an altitude of 8500 feet at 4:02 P.M. and 12,700 feet at 4:39 P.M. What was the average rate of change in the altitude in feet per minute?

_____

10. Enrollment in a college was 11,500 two years ago, 10,975 last year, and 10,800 this year. What is the average rate of change in enrollment per year for this 3-year period?

We use the ordered pairs (1, 11,500) and (3, 10,800).

$$\text{average rate of change} = \frac{10,800 - 11,500}{3 - 1}$$

$$= \frac{-700}{2}$$

$$= -350$$

The enrollment decreases at a rate of 350 students per year.

10. A company had 44 employees during the first year of operation. During their eighth year, the company had 79 employees. What was the average rate of change in the number of employees per year?

_____

Objective 5 Practice Exercises

For extra help, see Examples 9–10 on pages 155–156 of your text.

Solve each problem.

13. Suppose in 2005, the sales of a company were $1,625,000. In 2010, the company had sales of $2,250,000. Find the average rate of change in the sales per year.

13. _____

14. A state had a population of 755,000 in 2000 and a population of 809,000 in 2012. Find the average rate of change in population per year.

14. _____

15. Suppose a man's salary was $45,750 in 1995 and $60,000 in 2010. Find the average rate of change in the salary per year.

15. _____

Chapter 2 LINEAR EQUATIONS, GRAPHS, AND FUNCTIONS

2.3   Writing Equations of Lines

Learning Objectives
1   Write an equation of a line, given its slope and *y*-intercept.
2   Graph a line, using its slope and *y*-intercept.
3   Write an equation of a line, given its slope and a point on the line.
4   Write an equation of a line, given two points on the line.
5   Write equations of horizontal and vertical lines.
6   Write an equation of a line parallel or perpendicular to a given line.
7   Write an equation of a line that models real data.

Key Terms

Use the vocabulary terms listed below to complete each statement in exercises 1−3.

slope-intercept form   **point-slope form**   **standard form**

1. A linear equation in the form $y - y_1 = m(x - x_1)$ is written in

   _____.

2. A linear equation in the form $Ax + By = C$ is written in

   _____.

3. A linear equation in the form $y = mx + b$ is written in

   _____.

Objective 1   Write an equation of a line given its slope and *y*-intercept.

Video Examples

Review this example for Objective 1:

1. Write an equation of the line with slope $\frac{5}{7}$ and *y*-intercept (0, –6).

   Here, $m = \frac{5}{7}$ and $b = -6$, so we can write the following equation.
   $$y = mx + b$$
   $$y = \frac{5}{7}x + (-6), \text{ or } y = \frac{5}{7}x - 6$$

Now Try:

1. Write an equation of the line with slope $\frac{7}{9}$ and *y*-intercept (0, 8).

   _____

Name: _____ Date: _____

Instructor: _____ Section: _____

Objective 1 Practice Exercises

For extra help, see Example 1 on page 163 of your text.

Write the slope-intercept form equation of the line with the given slope and y-intercept.

1. $m = \dfrac{3}{2}$; $b = -\dfrac{2}{3}$

1. _____

2. $m = -7$; $b = -2$

2. _____

3.   Slope: $-\dfrac{6}{5}$; y-intercept $\left(0,\ \dfrac{2}{5}\right)$

3. _____

Objective 2   Graph a line, using its slope and y-intercept.

Video Examples

Review this example for Objective 2:

2.   Graph the equation by using the slope and y-intercept.

$$2x - 3y = 6$$

Solve for y to write the equation in slope-intercept form.

$$2x - 3y = 6$$
$$-3y = -2x + 6$$
$$y = \frac{2}{3}x - 2$$

The y-intercept is $(0, -2)$. Graph this point.

The slope is $\dfrac{2}{3}$. By definition,

$$\text{slope } m = \frac{\text{change in } y \text{ (rise)}}{\text{change in } x \text{ (run)}} = \frac{2}{3}$$

From the y-intercept, count up 2 units and to the right 3 units to obtain the point $(3, 0)$.

Draw the line through the points $(0, -2)$ and $(3, 0)$ to obtain the graph.

Now Try:

2.   Graph the equation by using the slope and y-intercept.

$$2x - 3y = 0$$

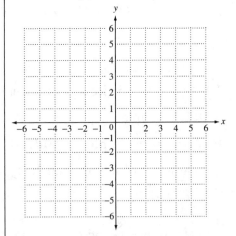

Name: Date:
Instructor: Section:

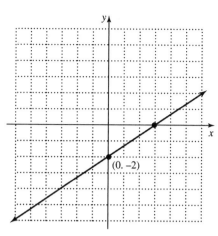

Objective 2 Practice Exercises

For extra help, see Example 2 on pages 163–164 of your text.

Graph each equation by using the slope and y-intercept.

4. $4x - y = 4$

4.

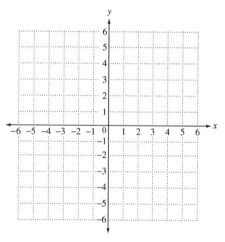

5. $y = -3x + 6$

5.

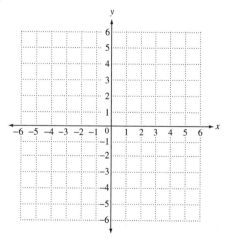

Graph the line passing through the given point and having the given slope.

6. $(-2,-2)$; $m = 0$

6.

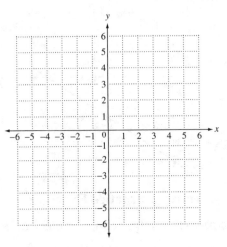

Objective 3   **Write an equation of a line, given its slope and a point on the line.**

Video Examples

Review this example for Objective 3:

3.   Write an equation of the line with slope $\frac{2}{3}$ passing through the point $(4,-7)$.

Method 1 Use point-slope form, with $(x_1, y_1) = (4,-7)$ and $m = \frac{2}{3}$.

$$y - y_1 = m(x - x_1)$$

$$y - (-7) = \frac{2}{3}(x - 4)$$

$$y + 7 = \frac{2}{3}(x - 4)$$

$$3y + 21 = 2x - 8$$

$$3y = 2x - 29$$

$$y = \frac{2}{3}x - \frac{29}{3}$$

Method 2 Use slope-intercept form, with $(x_1, y_1) = (4,-7)$ and $m = \frac{2}{3}$.

$$y = mx + b$$

$$-7 = \frac{2}{3}(4) + b$$

$$-7 = \frac{8}{3} + b$$

$$-\frac{29}{3} = b, \text{ or } b = -\frac{29}{3}$$

Now Try:

3.   Write an equation of the line with slope $\frac{4}{5}$ passing through the point $(6,-4)$.

_____

Knowing $m = \frac{2}{3}$ and $b = -\frac{29}{3}$ gives the

equation $y = \frac{2}{3}x - \frac{29}{3}$, same as Method 1.

Objective 3 Practice Exercises

For extra help, see Example 3 on page 165 of your text.

Write the equation in standard form of the line satisfying the given conditions.

7. $(-3, 4)$; $m = -\frac{3}{5}$

7. _____

8. $(-4, -7)$; $m = \frac{4}{3}$

8. _____

9. $(-1, 2)$; $m = \frac{2}{3}$

9. _____

Objective 4   Write an equation of a line, given two points on the line.

Video Examples

Review this example for Objective 4:

4. Write the equation of the line passing through the point (6, 8) and (–3, 5). Give the final answer in slope-intercept form and then in standard form.

First, find the slope of the line.

$(x_1, y_1) = (6,\ 8)$ and $(x_2, y_2) = (-3,\ 5)$

slope $m = \dfrac{y_2 - y_1}{x_2 - x_1} = \dfrac{5-8}{-3-6} = \dfrac{-3}{-9} = \dfrac{1}{3}$

Now Try:

4. Write the equation of the line passing through the point (7, 15) and (15, 9). Give the final answer in slope-intercept form and then in standard form.

_____

Now use (x_1, y_1), here $(6, 8)$ and point-slope form.

$$y - y_1 = m(x - x_1)$$

$$y - 8 = \frac{1}{3}(x - 6)$$

$$y - 8 = \frac{1}{3}x - 2$$

$$y = \frac{1}{3}x + 6 \quad \text{Slope-intercept form}$$

$$3y = x + 18$$

$$-x + 3y = 18$$

$$x - 3y = -18 \quad \text{Standard form}$$

Objective 4 Practice Exercises

For extra help, see Example 4 on page 166 of your text.

Write the equation in standard form of the line through the given points.

10. $(3, 7), (5, 4)$ **10.** _____

11. $(2, -1), (5, -2)$ **11.** _____

12. $(-1, -4), (-2, -3)$ **12.** _____

Objective 5   Write equations of horizontal and vertical lines.

Video Examples

Review these examples for Objective 5:

5. Write an equation of the line passing through the point $(2, -2)$ that satisfies the given condition.

a. The line has slope 0.

Since the slope is 0, this is a horizontal line.
$$y = -2.$$

Now Try:

5. Write an equation of the line passing through the point $(-5, 5)$ that satisfies the given condition.

a. The line has slope 0.

_____

b. The line has undefined slope.

This is a vertical line, since the slope is undefined.

$x = 2$

b. The line has undefined slope.

_____

Objective 5 Practice Exercises

For extra help, see Example 5 on page 167 of your text.

Write the equation in standard form of the line through the given points.

13. $(-1, -7), (-1, 8)$

13. _____

14. $(0, 2), (0, -6)$

14. _____

15. $(4, -5), (8, -5)$

15. _____

Objective 6   **Write an equation of a line parallel or perpendicular to a given line.**

Video Examples

Review these examples for Objective 6:

6. Write an equation in slope-intercept form of the line passing through the point $(-4, 5)$ that satisfies the given condition.

     a. The line is parallel to $5x + 2y = 10$.

First, find the slope of the given line.
$$5x + 2y = 10$$
$$2y = -5x + 10$$
$$y = -\frac{5}{2}x + 5$$

The slope is $-\frac{5}{2}$.

Use point-slope form with $(x_1, y_1) = (-4, 5)$ and $m = -\frac{5}{2}$.

Now Try:

6. Write an equation in slope-intercept form of the line passing through the point $(-6, 8)$ that satisfies the given condition.

     a. The line is parallel to $3x + 4y = 12$.

_____

$$y - y_1 = m(x - x_1)$$
$$y - 5 = -\frac{5}{2}(x - (-4))$$
$$y - 5 = -\frac{5}{2}(x + 4)$$
$$y - 5 = -\frac{5}{2}x - 10$$
$$y = -\frac{5}{2}x - 5$$

b. The line is perpendicular to $5x + 2y = 10$.

From part (a), the line in slope-intercept form is $y = -\frac{5}{2}x + 5$.

The line perpendicular to this line must have slope $\frac{2}{5}$, the negative reciprocal of $-\frac{5}{2}$.

Use point-slope form with $(x_1, y_1) = (-4, 5)$ and $m = \frac{2}{5}$.

$$y - y_1 = m(x - x_1)$$
$$y - 5 = \frac{2}{5}(x - (-4))$$
$$y - 5 = \frac{2}{5}(x + 4)$$
$$y - 5 = \frac{2}{5}x + \frac{8}{5}$$
$$y = \frac{2}{5}x + \frac{33}{5}$$

b. The line is perpendicular to $3x + 4y = 12$.

_____

Objective 6 Practice Exercises

For extra help, see Example 6 on pages 167–168 of your text.

Write the equation in standard form of the line satisfying the given conditions.

16.   parallel to $2x + 3y = -12$, through $(9, -3)$ **16.** _____

17.   parallel to $4x - 3y = 8$, through $(-2, 3)$.   **17.** _____

18.   perpendicular to $x - 3y = 0$, through $(-10, 2)$ **18.** _____

Name: Date:

Instructor: Section:

Objective 7   Write an equation of a line that models real data.

Video Examples

Review these examples for Objective 7:

8. The table and scatter graph shows the number of internet users in the world from 1998 to 2005, where year 0 represents 1998.

Year	Number of Internet Users (millions)
0	147
2	361
4	587
6	817
8	1093

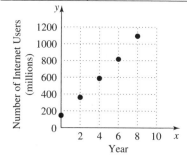

Now Try:

8. The table and scatter graph shows the average annual telephone expenditures for residential and pay telephones from 2001 to 2006, where year 0 represents 2001.

Year	Annual Telephone Expenditures
0	$686
2	$620
3	$592
4	$570
5	$542

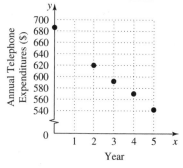

a. Find an equation that models the data.

The points appear to lie approximately in a straight line. y represents the number of internet users in year x. To find an equation of the line, we choose the ordered pairs (0, 147) and (8, 1093) from the table and find the slope of the line through these points.

$$(x_1, y_1) = (0,\ 147) \text{ and } (x_2, y_2) = (8,\ 1093)$$

$$\text{slope } m = \frac{y_2 - y_1}{x_2 - x_1} = \frac{1093 - 147}{8 - 0} = \frac{946}{8}$$

$$= 118.25$$

a. Find an equation that models the data.

_____

Use the slope, 118.25, and the point (0, 147) in slope-intercept form.

$$y = mx + b$$

$$147 = 118.25(0) + b$$

$$147 = b$$

Thus, $m = 118.25$ and $b = 147$, so the equation of the line is $y = 118.25x + 147$.

b. Find and interpret the ordered pair associated with the equation for $x = 5$.

If $x = 5$, then

$$y = 118.25(5) + 147$$

$$= 591.25 + 147$$

$$= 738.25$$

In 2003, there were 738.25 million internet users.

b. Find the ordered pair associated with the equation for $x = 1$.

_____

Objective 7 Practice Exercises

For extra help, see Examples 7–9 on pages 169–171 of your text.

Solve each problem.

19. To run a newspaper ad, there is a $25 set up fee plus a charge of $1.25 per line of type in the ad. Let x represent the number of lines in the ad so that y represents the total cost of the ad (in dollars).
a. Write an equation in the form $y = mx + b$.
b. Give three ordered pairs associated with the equation for x-values 0, 5, and 10.

19.

a. _____

b. _____

20. The table and scatter graph shows the U.S. municipal solid waste recycling percent since 1985, where year 0 represents 1985.
a. Find an equation that models the data.
b. Use the equation from part (a) to predict the percent of municipal solid waste recycling in the year 2015.

20.

a. _____

b. _____

Year	Recycling Percent
0	10.1
5	16.2
10	26.0
15	29.1
20	32.5

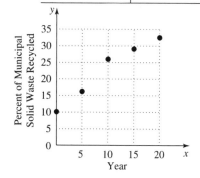

Chapter 2 LINEAR EQUATIONS, GRAPHS, AND FUNCTIONS

2.4   Linear Inequalities in Two Variables

Learning Objectives
1   Graph linear inequalities in two variables.
2   Graph the intersection of two linear inequalities.
3   Graph the union of two linear inequalities.

Key Terms

Use the vocabulary terms listed below to complete each statement in exercises 1–2.

linear inequality in two variables   **boundary line**

1.   In the graph of a linear inequality, the _____
separates the region that satisfies the inequality from the region that does not
satisfy the inequality.

2.   An inequality that can be written in the form $Ax + By < C$, $Ax + By > C$,
$Ax + By \leq C$, or $Ax + By \geq C$ is called a_____.

Objective 1   Graph linear inequalities in two variables.

Video Examples

Review these examples for Objective 1:

1.   Graph $3x - 2y \leq 6$.

The inequality $3x - 2y \leq 6$ means that
$3x - 2y < 6$ or $3x - 2y = 6$.
We begin by graphing the line $3x - 2y = 6$ with
intercepts (0, –3) and (2, 0). This boundary line
divides the plane into two regions, one of which
satisfies the inequality. We use the test point
(0, 0) to see whether the resulting statement is
true or false, thereby determining whether the
point is in the shaded region or not.

$$3x - 2y \leq 6$$
$$3(0) - 2(0) \overset{?}{\leq} 6$$
$$0 - 0 \overset{?}{\leq} 6$$
$$0 \leq 6 \quad \text{True}$$

Since the last statement is true, we shade the
region that includes the test point (0, 0). The
shaded region, along with the boundary line, is
the desired graph.

Now Try:

1.   Graph $2x + 5y \leq -8$.

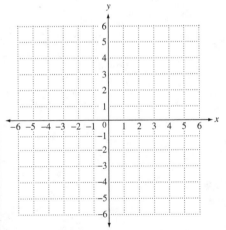

Name: Date:
Instructor: Section:

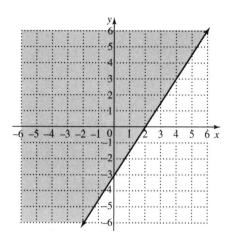

2. Graph $y \geq 3x$.

We graph $y = 3x$ using a solid line through
$(0, 0)$, $(1, 3)$ and $(2, 6)$. Because $(0, 0)$ is on the
line $y \geq 3x$, it cannot be used as a test point.
Instead we choose a test point off the line, say
$(3, 0)$.

$0 \overset{?}{\geq} 3(3)$

$0 \geq 9$ False

Because $0 \geq 9$ is false, shade the other region.

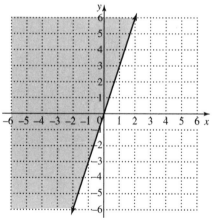

2. Graph $y \geq x$.

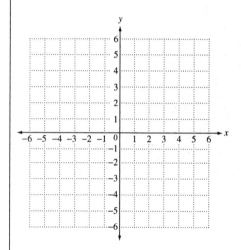

Objective 1 Practice Exercises

For extra help, see Examples 1–3 on pages 180–182 of your text.

Graph each linear inequality.

1. $x - y < 5$

1.

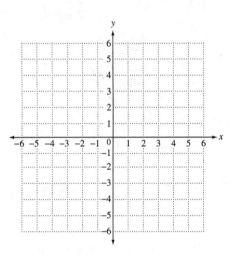

2. $2x + 3y \geq 6$

2.

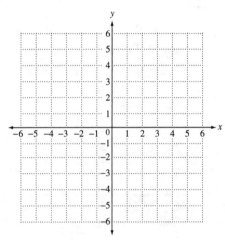

3. $x \leq 4y$

3.

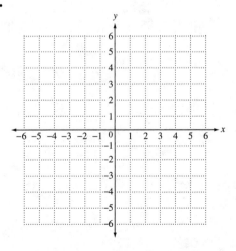

Objective 2   Graph the intersection of two linear inequalities.

Video Examples

Review this example for Objective 2:

4. Graph $x + y \leq 4$ and $x > 2$.

A pair of inequalities joined with the word *and* is interpreted as the intersection of the solution sets of the inequalities. The graph of the intersection of two or more inequalities is the region of the plane where all points satisfy all of the inequalities at the same time.

Begin by graphing each of the two inequalities separately.

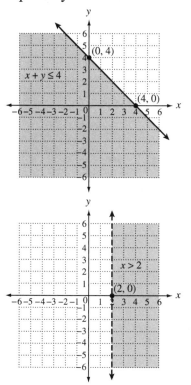

The intersection of the two graphs is the graph we are seeking.

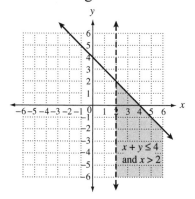

Now Try:

4. Graph $3x - 4y < 12$ and $y > -4$.

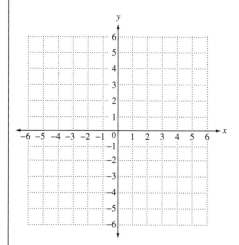

Name: Date:

Instructor: Section:

Objective 2 Practice Exercises

For extra help, see Example 4 on pages 182–183 of your text.

Graph the intersection of each pair of linear inequalities.

4. $x - y < 3$ and $x + y > -2$

4.

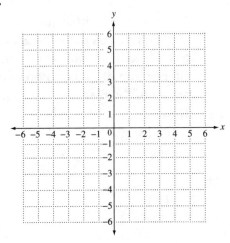

5. $4x + y \leq 4$ and $x - 2y \leq -2$

5.

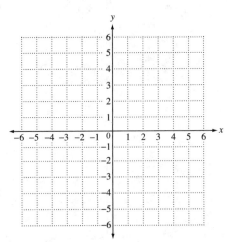

6. $x > 2$ and $2x - 3y < 6$

6.

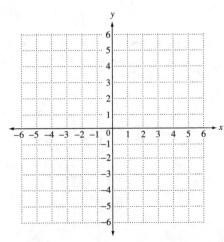

Objective 3   Graph the union of two linear inequalities.

Video Examples

Review this example for Objective 3:

5. Graph $x + y \le 4$ or $x > 2$.

When two inequalities are joined by the word *or*, we must find the union of the graphs of the inequalities. The graph of the union of two inequalities includes all of the points that satisfy either inequality.

Begin by graphing each of the two inequalities separately.

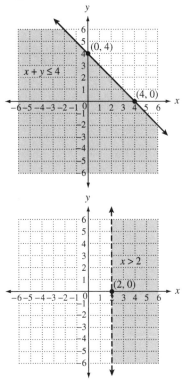

The union of the two graphs is the graph we are seeking.

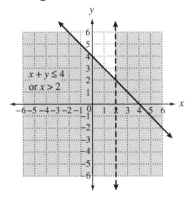

Now Try:

5. Graph $4x + 3y < 12$ or $x \ge 2$.

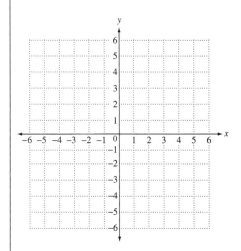

Objective 3 Practice Exercises

For extra help, see Example 5 on page 183 of your text.

Graph the union of each pair of linear inequalities.

7. $4x - 2y \geq -4$ or $x \geq 1$ **7.**

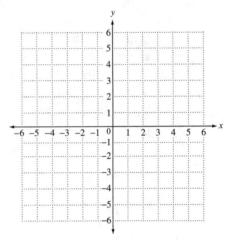

8. $x \geq 4$ or $y < -3$ **8.**

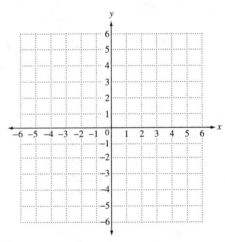

9. $x + y \geq 0$ or $x - y \geq 0$ **9.**

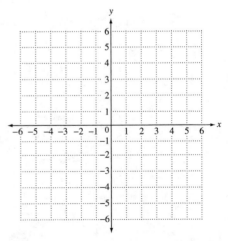

Chapter 2 LINEAR EQUATIONS, GRAPHS, AND FUNCTIONS

2.5   Introduction to Relations and Functions

Learning Objectives
1   Define and identify relations and functions.
2   Find domain and range.
3   Identify functions defined by graphs and equations.

Key Terms

Use the vocabulary terms listed below to complete each statement in exercises 1−6.

dependent variable	independent variable	relation
function	domain	range

1.   The _____ of a relation is the set of second components (y-values) of the ordered pairs of the relation.

2.   A _____ is a set of ordered pairs of real numbers.

3.   If the quantity y depends on x, then y is called the _____ in a relation between x and y.

4.   The _____ of a relation is the set of first components (x-values) of the ordered pairs of the relation.

5.   A _____ is a set of ordered pairs in which each value of the first component, x, corresponds to exactly one value of the second component, y.

6.   If the quantity y depends on x, then x is called the _____ in a relation between x and y.

Objective 1   Define and identify relations and functions.

Video Examples

Review these examples for Objective 1:

1.   Write the relation as a set of ordered pairs.

Number of Hours Worked	Paycheck Amount (in dollars)
8	96
16	192
24	288
32	384

{(8, 96), (16, 192), (24, 288), (32, 384)}

Now Try:

1.   Write the relation as a set of ordered pairs.

Number of Hours Fishing	Number of Fish Caught
1	3
2	4
3	6
5	7

_____

2. Determine whether each relation defines a function.

   a. $G = \{(-5, -2), (-2, 6), (6, 8), (8, 11),$
   $(11, 11)\}$

   Relation G is a function. Although the last two ordered pairs have the same y-value, this does not violate the definition of a function.

   b. $H = \{(-9, 2), (-6, 2), (-6, 9)\}$

   In relation H, the last two ordered pairs have the same x-value pair with different y-values. H is a relation, but not a function.

2. Determine whether each relation defines a function.

   a. $\{(10, 1), (100, 2), (70, 2)\}$

   _____

   b. $\{(0, 2), (2, 6), (6, 3), (0, 7)\}$

   _____

Objective 1 Practice Exercises

For extra help, see Examples 1–2 on pages 186–187 of your text.

Write the relation as a set of ordered pairs.

1.

x	y
1	3
1	4
2	−1
3	7

1. _____

Decide whether each relation is a function.

2. $\{(2, -2,), (3, -3), (4, -4)\}$

2. _____

3. $\{(3, 4), (5, 2), (4, 3), (5, 3), (-2, 2)\}$

3. _____

Objective 2   Find domain and range.

Video Examples

Review these examples for Objective 2:

3. Give the domain and range of each relation. Tell whether the relation defines a function.

   $\{(15, 2), (20, 3), (6, 10), (-1, 2)\}$

   The domain is the set of x-values $\{-1, 6, 15, 20\}$. The range is the set of y-values $\{2, 3, 10\}$. The relation is a function, because each x-value corresponds to exactly one y-value.

Now Try:

3. Give the domain and range of each relation. Tell whether the relation defines a function.
   $\{(13, -1), (13, -2), (13, 4)\}$

   domain: _____

   range: _____

   _____

4. Give the domain and range of each relation.

a.

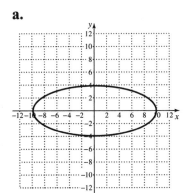

The *x*-values of the points on the graph include all numbers between –10 and 10, inclusive. The *y*-values include all numbers between –4 and 4, inclusive.

The domain is [–10, 10]. The range is [–4, 4].

b.

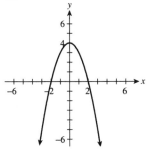

The graph extends indefinitely left and right, as well as downward. The domain is $(-\infty, \infty)$. Because there is a greatest *y*-value, 4, the range includes all numbers less than or equal to 4, written $(-\infty, 4]$.

4. Give the domain and range of each relation.

a.

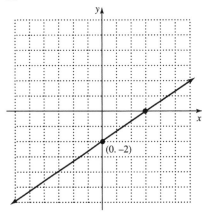

(0. –2)

_____

_____

b.

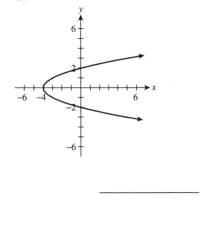

_____

_____

Objective 2 Practice Exercises

For extra help, see Examples 3–4 on pages 189–190 of your text.

Decide whether the relation is a function, and give the domain and range of the relation.

4. $\{(5,\ 2),\ (3,-1),\ (1,-3),\ (-1,-5)\}$

4. _____

domain: _____

range: _____

5.

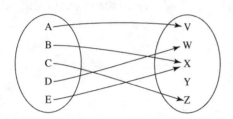

5. _____

domain:_____

range: _____

6.

x	y
1	3
2	−1
−1	4
1	4

6. _____

domain:_____

range: _____

Objective 3   Identify functions defined by graphs and equations.

Video Examples

Review these examples for Objective 3:

5. Use the vertical line test to determine whether the relation graphed is a function.

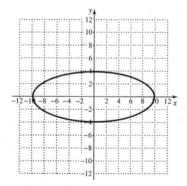

The graph is not a function.

6. Decide whether the relation defines y as a function of x. Give the domain.

$$y = 2x - 4$$

Each x value corresponds to just one y-value and the relation defines a function. Since x can be any real number, the domain is $(-\infty, \infty)$.

Now Try:

5. Use the vertical line test to determine whether the relation graphed is a function.

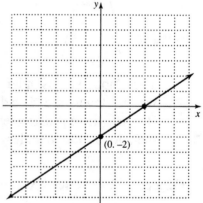

(0, −2)

_____

6. Decide whether the relation defines y as a function of x. Give the domain.

$$y = 3x - 1$$

_____

_____

Name: Date:

Instructor: Section:

Objective 3 Practice Exercises

For extra help, see Examples 5–6 on pages 191–192 of your text.

Use the vertical line test to determine whether the relation graphed is a function.

7.   7. _____

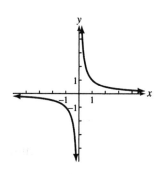

Decide whether each equation defines y as a function of x. Give the domain.

8. $y^2 = x + 1$ 8. _____

9. $y = \dfrac{3}{x+6}$ 9. _____

Chapter 2 LINEAR EQUATIONS, GRAPHS, AND FUNCTIONS

2.6   Function Notation and Linear Functions

Learning Objectives
1   Use function notation.
2   Graph linear and constant functions.

Key Terms

Use the vocabulary terms listed below to complete each statement in exercises 1–3.

function notation   linear function   constant function

1.   A function defined by an equation of the form $f(x) = ax + b,$ for real numbers a and b, is a _____.

2.   _____ $f(x)$ represents the value of the function at x, that is, the y-value that corresponds to x.

3.   A _____ is a linear function of the form $f(x) = b,$ for a real number b.

Objective 1   Use function notation.

Video Examples

Review these examples for Objective 1:

1.   Let $f(x) = 7x - 3$. Evaluate the function f for the following.

   $x = 4$

   Start with the given function. Replace x with 4.
   $$f(x) = 7x - 3$$
   $$f(4) = 7(4) - 3$$
   $$f(4) = 28 - 3$$
   $$f(4) = 25$$

   Thus, $f(4) = 25$.

3.   Let $g(x) = 5x + 6$. Find and simplify $g(n+8)$.

   Replace x with $n + 8$.
   $$g(x) = 5x + 6$$
   $$g(n+8) = 5(n+8) + 6$$
   $$g(n+8) = 5n + 40 + 6$$
   $$g(n+8) = 5n + 46$$

Now Try:

1.   Let $f(x) = 8x - 7$. Evaluate the function f for the following.

   $x = 3$

   _____

3.   Let $g(x) = 4x - 7$. Find and simplify $g(a-1)$.

   _____

4. For the function, find $f(5)$.

$f = \{(7,-27),\ (5,-25),\ (3,-23),\ (1,-21)\}$

From the ordered pair (5, –25), we have $f(5) = -25$.

6. Write the equation using function notation $f(x)$. Then find $f(-5)$.

$x - 5y = 8$

Step 1 $\quad x - 5y = 8$

$\qquad\qquad -5y = -x + 8$

$\qquad\qquad y = \dfrac{1}{5}x - \dfrac{8}{5}$

Step 2 $\quad f(x) = \dfrac{1}{5}x - \dfrac{8}{5}$

$\qquad\qquad f(-5) = \dfrac{1}{5}(-5) - \dfrac{8}{5}$

$\qquad\qquad f(-5) = -\dfrac{13}{5}$

4. For the function, find $f(-6)$.

$f = \{(-2,\ 11),\ (-4,\ 17),$
$\qquad (-6,\ 21),\ (-8,\ 24)\}$

_____

6. Write the equation using function notation $f(x)$. Then find $f(-3)$.
$2x + 3y = 7$

_____

Objective 1 Practice Exercises

For extra help, see Examples 1–6 on pages 197–200 of your text.

For each function f, find (a) $f(-2)$, *(b)* $f(0)$, *and (c)* $f(-x)$.

1. $\quad f(x) = 3x - 7$

1. a. _____

b. _____

c. _____

2. $\quad f(x) = 2x^2 + x - 5$

2. a. _____

b. _____

c. _____

3. $\quad f(x) = 9$

3. a. _____

b. _____

c. _____

Objective 2   Graph linear and constant functions.

Video Examples

Review this example for Objective 2:

7. Graph the function $f(x) = -2x - 3$. Give the domain and range.

The graph of the function has slope -2 and y-intercept -3. To graph this function, plot the y-intercept $(0, -3)$ and use the definition of slope as $\frac{\text{rise}}{\text{run}}$ to find a second point on the line. Since the slope is -2, move down two units and right one unit to the point $(1, -5)$. Draw the straight line through the points to obtain the graph. The domain and range are both $(-\infty, \infty)$.

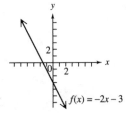

Now Try:

7. Graph the function $f(x) = \frac{1}{2}x + \frac{1}{2}$. Give the domain and range.

_____

_____

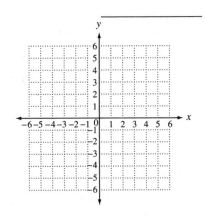

Objective 2 Practice Exercises

For extra help, see Example 7 on page 201 of your text.

Graph each function. Give the domain and range.

4. $2x - y = -2$

4. domain _____

range _____

5. $y + \dfrac{1}{2}x = -2$

5. domain _____

   range _____

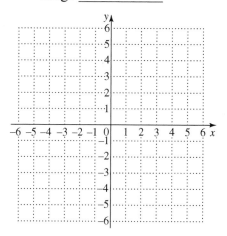

6. $y = 2$

6. domain _____

   range _____

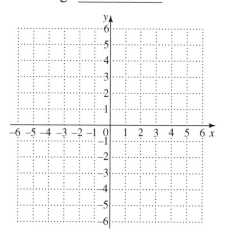

Chapter 3 SYSTEMS OF LINEAR EQUATIONS

3.1   Systems of Linear Equations in Two Variables

Learning Objectives	
1	Decide whether an ordered pair is a solution of a linear system.
2	Solve linear systems by graphing.
3	Solve linear systems (with two equations and two variables) by substitution.
4	Solve linear systems (with two equations and two variables) by elimination.
5	Solve special systems.

Key Terms

Use the vocabulary terms listed below to complete each statement in exercises 1−7.

> **system of equations**   **linear system**
>
> **solution set of a system**   **consistent system**   **inconsistent system**
>
> **independent equations**   **dependent equations**

1. Equations of a system that have different graphs are called
    _____.

2. A system of equations with at least one solution is a
    _____.

3. Two or more equations that are to be solved at the same time form a
    _____.

4. The _____ of linear equations includes all
    the ordered pairs that make all the equations of the system true at the same time.

5. Equations of a system that have the same graph (because they are different forms
    of the same equation) are called _____.

6. A system with no solution is called a(n) _____.

7. A(n) _____ consists of two or more linear
    equations with the same variables.

Objective 1   Decide whether an ordered pair is a solution of a linear system.

Video Examples

Review this example for Objective 1:

1. Decide whether the given ordered pair is a solution of the given system.

$$x + y = 8; \quad (3, 5)$$
$$2x - y = 1$$

Replace x with 3 and y with 5 in each equation of the system.

$$
\begin{array}{c|c}
x + y = 8 & 2x - y = 1 \\
\overset{?}{3 + 5 = 8} & \overset{?}{2(3) - 5 = 1} \\
\text{True} \quad 8 = 8 & \text{True} \quad 1 = 1
\end{array}
$$

Since (3, 5) makes both equations true, (3, 5) is a solution of the system.

Now Try:

1. Decide whether the given ordered pair is a solution of the given system.
$$x - 5y = 11; \quad (6, -1)$$
$$3x + 8y = 10$$

_____

Objective 1 Practice Exercises

For extra help, see Example 1 on pages 216–217 of your text.

Decide whether the given ordered pair is a solution of the given system. Write solution *or* not a solution.

1.   (4, 1)

     $2x + 3y = 11$

     $3x - 2y = 9$

1. _____

2. $(-3, -1)$

     $5x - 3y = -12$

     $2x + 3y = -9$

2. _____

3.   (4, 0)

     $4x + 3y = 16$

     $x - 4y = -4$

3. _____

Name: Date:

Instructor: Section:

Objective 2   Solve a linear system by graphing.

Video Examples

Review this example for Objective 2:

2.   Solve the system of equations by graphing.

$$x - 2y = 6 \quad (1)$$

$$2x + y = 2 \quad (2)$$

To graph these linear equations, we plot several points for each line.

$x - 2y = 6$ 　　　　　$2x + y = 2$

x	y
0	-3
6	0
4	-1

x	y
0	2
1	0
-1	4

From the graph we see that the solution is $(2,-2)$.

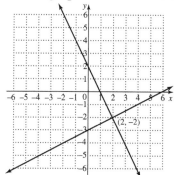

To check, substitute 2 for x and -2 for y in each equation.

$x - 2y = 6$	$2x + y = 2$
$\overset{?}{2 - 2(-2) = 6}$	$\overset{?}{2(2) + (-2) = 2}$
$\overset{?}{2 + 4 = 6}$	$\overset{?}{4 - 2 = 2}$
True $6 = 6$	True $2 = 2$

The solution set is $\{(2,-2)\}$.

Now Try:

2.   Solve the system of equations by graphing.

$$2x - 5y = 8 \quad (1)$$

$$5x - 4y = 3 \quad (2)$$

_____

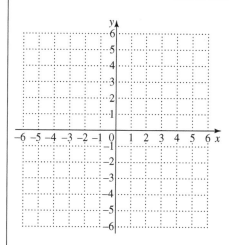

Name: Date:

Instructor: Section:

Objective 2 Practice Exercises

For extra help, see Example 2 on page 217 of your text.

Solve the system by graphing.

4. $2x + 3y = 5$

    $3x - y = 13$

4. _____

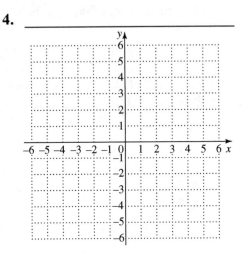

5. $2x = y$

    $5x + 3y = 0$

5. _____

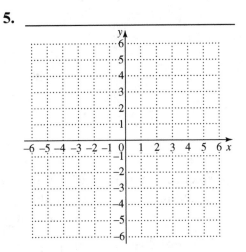

6. $y - 2 = 0$

    $3x - 4y = -17$

6. _____

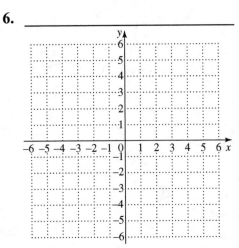

Objective 3   Solve linear systems (with two equations and two variables) by substitution.

Video Examples

Review this example for Objective 3:

4. Solve the system by substitution.

$$3x - 2y = 16 \quad (1)$$
$$x - 7y = -1 \quad (2)$$

Step 1 First, solve equation (2) for x.
$$x - 7y = -1$$
$$x = 7y - 1$$

Step 2 Substitute $7y - 1$ for x in equation (1).
$$3x - 2y = 16$$
$$3(7y - 1) - 2y = 16$$

Step 3 Solve for y.
$$21y - 3 - 2y = 16$$
$$19y - 3 = 16$$
$$19y = 19$$
$$y = 1$$

Step 4 Find x. From Step 1, $x = 7y - 1$.
Substitute 1 for y.
$$x = 7(1) - 1 = 6$$

Step 5 Check the solution (6, 1) in both equations (1) and (2).

$$
\begin{array}{c|c}
3x - 2y = 16 & x - 7y = -1 \\
\quad\ ? & \quad\ ? \\
3(6) - 2(1) = 16 & 6 - 7(1) = -1 \\
\quad\ ? & \quad\ ? \\
18 - 2 = 16 & 6 - 7 = -1 \\
16 = 16 \ \text{True} & -1 = -1 \ \text{True}
\end{array}
$$

The solution set is $\{(6, 1)\}$.

Now Try:

4. Solve the system by substitution.

$$4x + 9y = 11 \quad (1)$$
$$2x - y = 11 \quad (2)$$

_____

Objective 3 Practice Exercises

For extra help, see Examples 3–5 on pages 218–221 of your text.

Solve each system by substitution.

7. $3x - 2y = -1$

    $x = \dfrac{3}{4}y$

7. _____

8. $2x + y = 6$

    $y = 5 - 3x$

8. _____

9. $y = 11 - 2x$

    $x = 18 - 3y$

9. _____

Objective 4   Solve linear systems (with two equations and two variables) by elimination.

Video Examples

Review this example for Objective 4:

7. Solve the system by elimination.

    $4x + 3y = 41$ (1)

    $3x - 2y = 1$ (2)

    Step 1 Both equations are in standard form.

    Step 2 Multiply equation (1) by 2 and equation (2) by 3.

    $8x + 6y = 82$ Multiply (1) by 2.

    $9x - 6y = 3$ Multiply (2) by 3.

Now Try:

7. Solve the system by elimination.

    $4x + 5y = 11$ (1)

    $3x - 2y = -9$ (2)

    _____

Step 3 Now add.

$$8x + 6y = 82$$
$$9x - 6y = 3$$
$$\overline{17x = 85}$$

Step 4 Solve for *x*. $x = 5$ *Step 5*

To find *y*, substitute 5 for *x* in either equation (1) or (2).

$$3x - 2y = 1 \quad (2)$$
$$3(5) - 2y = 1$$
$$15 - 2y = 1$$
$$-2y = -14$$
$$y = 7$$

Step 6 Check the solution (5, 7) in both equations (1) and (2).

$$\begin{array}{l|l}
4x + 3y = 41 & 3x - 2y = 1 \\
? & ? \\
4(5) + 3(7) = 41 & 3(5) - 2(7) = 1 \\
? & ? \\
20 + 21 = 41 & 15 - 14 = 1 \\
41 = 41 \ \text{True} & 1 = 1 \ \text{True}
\end{array}$$

The solution set is $\{(5, 7)\}$.

Objective 4 Practice Exercises

For extra help, see Examples 6–7 on pages 221–223 of your text.

Solve each system by elimination.

10. $3x - y = 11$

  $x + y = 5$

10. _____

11. $\dfrac{1}{2}x + \dfrac{1}{4}y = 5$

  $\dfrac{1}{2}x - \dfrac{3}{4}y = -3$

11. _____

12. $x + 2y = 7$

$x - y = -2$

12. _____

Objective 5   Solve special systems.

Video Examples

Review these examples for Objective 5:

8. Solve the system.

$x + 5y = 9$ (1)

$3x + 15y = 27$ (2)

We multiply equation (1) by –3 and then add the result to equation (2).

$-3x - 15y = -27$ Multiply (1) by -3.

$\underline{3x + 15y = 27}$ (2)

$0 = 0$ True

The equations are dependent.
Since the statement is true, the solution set is the set of all points on the line with equation $x + 5y = 9$, written as $\{(x, y) \mid x + 5y = 9\}$.

9. Solve the system.

$5x + 5y = 9$ (1)

$x + y = 3$ (2)

We multiply equation (2) by –5 and then add the result to equation (1).

$5x + 5y = 9$ (1)

$\underline{-5x - 5y = -15}$ Multiply (2) by -5.

$0 = -6$ False

The system is inconsistent.
Since the statement is false, there are no ordered pairs that satisfy both equations.
The solution set is $\varnothing$.

Now Try:

8. Solve the system.

$12x + 9y = 36$ (1)

$4x + 3y = 12$ (2)

_____

9. Solve the system.

$4x + 3y = 12$ (1)

$8x + 6y = -24$ (2)

_____

10. Write the pair of equations in slope-intercept form, and use the results to tell how many solutions the system has.

$$5x - y = 4$$
$$15x - 3y = 12$$

Solve each equation for y.

$5x - y = 4$	$15x - 3y = 12$
$-y = -5x + 4$	$5x - y = 4$
$y = 5x - 4$	Same result

The lines have the same slope and same y-intercept, indicating that they coincide. There are infinitely many solutions.

10. Write the pair of equations in slope-intercept form, and use the results to tell how many solutions the system has.

$$2x - 5y = 7$$
$$6x - 15y = 21$$

_____

_____

_____

Objective 5 Practice Exercises

For extra help, see Examples 8–10 on pages 223–225 of your text.

Solve each system of equations using any method.

13. $8x + 4y = -1$
         $4x + 2y = 3$

13. _____

14. $x + 2y = 4$
         $8y = -4x + 16$

14. _____

15. $-3x + 2y = 6$
         $-6x + 4y = 12$

15. _____

Chapter 3 SYSTEMS OF LINEAR EQUATIONS

3.2 Systems of Linear Equations in Three Variables

Learning Objectives
1	Understand the geometry of systems of three equations in three variables.
2	Solve linear systems (with three equations and three variables) by elimination.
3	Solve linear systems (with three equations and three variables) in which some of the equations have missing terms.
4	Solve special systems.

Key Terms

Use the vocabulary terms listed below to complete each statement in exercises 1–3.

ordered triple   inconsistent system   dependent system

1. The solution of a linear system of equations in three variables is written as a(n) _____.

2. A system of equations in which all solutions of the first equation are also solutions of the second equation is a(n) _____.

3. A system of equations that has no common solution is called a(n) _____.

Objective 1 Understand the geometry of systems of three equations in three variables.

Objective 1 Practice Exercises

For extra help, see pages 232–233 of your text.

Answer each question.

1. If a system of linear equations in three variables has a single solution, how do the planes that are the graphs of the equations intersect?

1. _____

2. If a system of linear equations in three variables has no solution, how do the planes that are the graphs of the equations intersect?

2. _____

Objective 2 **Solve linear systems (with three equations and three variables) by elimination.**

Video Examples

Review this example for Objective 2:

1. Solve the system.

$$x - 2y + 5z = -7 \quad (1)$$
$$2x + 3y - 4z = 14 \quad (2)$$
$$3x - 5y + z = 7 \quad (3)$$

Step 1: Since x in equation (1) has coefficient 1, choose x as the focus variable and (1) as the working equation.

Step 2: Multiply working equation (1) by -2 and add the result to equation (2) to eliminate focus variable x.

$$-2x + 4y - 10z = 14 \quad \text{Multiply (1) by } -2$$
$$\underline{2x + 3y - 4z = 14 \quad (2)}$$
$$7y - 14z = 28 \quad (4)$$

Step 3: Multiply working equation (1) by -3 and add the result to equation (3) to eliminate focus variable x.

$$-3x + 6y - 15z = 21 \quad \text{Multiply (1) by } -3$$
$$\underline{3x - 5y + z = 7 \quad (3)}$$
$$y - 14z = 28 \quad (5)$$

Step 4: Write equations (4) and (5) as a system, then solve the system.

$$7y - 14z = 28 \quad (4)$$
$$y - 14z = 28 \quad (5)$$

We will eliminate z.

$$-7y + 14z = -28 \quad \text{Multiply (4) by } -1$$
$$\underline{y - 14z = 28 \quad (5)}$$
$$-6y = 0 \quad \text{Add.}$$
$$y = 0 \quad \text{Divide by } -6.$$

Substitute 0 for y in either equation to find z.

$$y - 14z = 28 \quad (5)$$
$$0 - 14z = 28 \quad \text{Let } y = 0.$$
$$-14z = 28$$
$$z = -2$$

Now Try:

1. Solve the system.

$$2x + y + 2z = -1$$
$$3x - y + 2z = -6$$
$$3x + y - z = -10$$

_____

Step 5: Now substitute $y = 0$ and $z = -2$ in working equation (1) to find the value of the remaining variable, focus variable x.

$$x - 2y + 5z = -7 \quad (1)$$
$$x - 2(0) + 5(-2) = -7 \quad \text{Let } y = 0 \text{ and } z = -2.$$
$$x - 10 = -7$$
$$x = 3$$

Step 6: It appears that the ordered triple $(3, 0, -2)$ is the solution of the system. We must check that the solution satisfies all three original equations of the system.

$$x - 2y + 5z = -7 \quad (1)$$
$$3 - 2(0) + 5(-2) \overset{?}{=} -7$$
$$3 - 0 - 10 \overset{?}{=} -7$$
$$-7 = -7$$

Because $(3, 0, -2)$ also satisfies equations (2) and (3), the solution set is $\{(3, 0, -2)\}$.

Objective 2 Practice Exercises

For extra help, see Example 1 on pages 234–235 of your text.

Solve each system of equations.

3. $\begin{aligned} x + y + z &= 2 \\ x - y + z &= -2 \\ x - y - z &= -4 \end{aligned}$

3. _____

4. $\begin{aligned} 2x + y - z &= 9 \\ x + 2y + z &= 3 \\ 3x + 3y - z &= 14 \end{aligned}$

4. _____

5. $2x - 5y + 2z = 30$

$x + 4y + 5z = -7$

$\frac{1}{2}x - \frac{1}{4}y + z = 4$

5. _____

Objective 3   Solve linear systems (with three equations and three variables) in which some of the equations have missing terms.

Video Examples

Review this example for Objective 3:

2.   Solve the system.

$3x \quad\quad - 4z = -23$ (1)

$y + 5z = 24$ (2)

$x - 3y \quad\quad = 2$ (3)

Since equation (1) is missing the variable y, one way to begin is to eliminate y again, using equations (2) and (3).

$3y + 15z = 72$ Multiply (2) by 3

$\underline{x - 3y \quad\quad = 2 \quad (3)}$

$x \quad\quad + 15z = 74$ (4)

Now solve the system composed of equations (1) and (4).

$3x - 4z = -23$ (1)

$\underline{-3x - 45z = -222}$ Multiply (4) by -3

$-49z = -245$

$z = 5$

Substitute 5 for z in (1) and solve for x.

$3x - 4z = -23$ (1)

$3x - 4(5) = -23$ Let $z = 5$.

$3x - 20 = -23$

$3x = -3$

$x = -1$

Now Try:

2.   Solve the system.

$x + 5y \quad\quad = -23$

$4y - 3z = -29$

$2x \quad\quad + 5z = 19$

_____

Substitute 5 for z in (2) and solve for y.

$$y + 5z = 24 \quad (2)$$
$$y + 5(5) = 24 \quad \text{Let } z = 5.$$
$$y + 25 = 24$$
$$y = -1$$

A check verifies that the solution set is $\{(-1, -1, 5)\}$.

Objective 3 Practice Exercises

For extra help, see Example 2 on pages 235–236 of your text.

Solve each system of equations.

6.
$$\begin{aligned} 7x + z &= -1 \\ 3y - 2z &= 8 \\ 5x + y \phantom{{}-2z} &= 2 \end{aligned}$$

6. _____

7.
$$\begin{aligned} 2x + 5y \phantom{{}+2z} &= 18 \\ 3y + 2z &= 4 \\ \tfrac{1}{4}x - y \phantom{{}+2z} &= -1 \end{aligned}$$

7. _____

8.
$$\begin{aligned} 5x \phantom{{}+4y} - 2z &= 8 \\ 4y + 3z &= -9 \\ \tfrac{1}{2}x + \tfrac{2}{3}y \phantom{{}+3z} &= -1 \end{aligned}$$

8. _____

Name: _____ Date: _____

Instructor: _____ Section: _____

Objective 4　Solve special systems.

Video Examples

Review these examples for Objective 4:

4.　Solve the system.

$$3x - 2y + 5z = 4 \quad (1)$$
$$-6x + 4y - 10z = -8 \quad (2)$$
$$\frac{3}{2}x - y + \frac{5}{2}z = 2 \quad (3)$$

Multiplying each side of equation (1) by -2 gives equation (2). Multiplying each side of equation (1) by $\frac{1}{2}$ gives equation (3). Thus, the equations are dependent, and all three equations have the same graph as shown below. The solution set is written

$$\{(x, y, z) \mid 3x - 2y + 5z = 4\}.$$

(1), (2), (3)

3.　Solve the system.

$$x - y + z = 7 \quad (1)$$
$$2x + 5y - 4z = 2 \quad (2)$$
$$-x + y - z = 4 \quad (3)$$

Since x in equation (1) has coefficient 1, choose x as the focus variable and (1) as the working equation. Using equations (1) and (3), we have

$$
\begin{array}{r}
x - y + z = 7 \quad (1) \\
-x + y - z = 4 \quad (3) \\
\hline
0 = 11 \quad \text{False}
\end{array}
$$

The resulting false statement indicates that equations (1) and (3) have no common solution. Thus, the system is inconsistent and the solution set is $\varnothing$. The graph of this system would show that three planes are parallel to each other as shown below.

(1)
(2)
(3)

Now Try:

4.　Solve the system.

$$x - 5y + 2z = 0$$
$$-x + 5y - 2z = 0$$
$$\frac{1}{2}x - \frac{5}{2}y + z = 0$$

_____

3.　Solve the system.

$$-4x - 2y + z = -19$$
$$-6x + 2y - 6z = -8$$
$$-4x + 2y - 5z = -6$$

_____

Objective 4 Practice Exercises

For extra help, see Examples 3–5 on pages 236–237 of your text.

Solve each system of equations.

9. $8x - 7y + 2z = 1$

       $3x + 4y - \ z = 6$

    $-8x + 7y - 2z = 5$

9. _____

10. $3x - 2y + 4z = \ 5$

    $-3x + 2y - 4z = -5$

    $\dfrac{3}{2}x - \ y + 2z = \ \dfrac{5}{2}$

10. _____

11. $-x \ + 5y - 2z = \ 3$

    $2x - 10y + 4z = -6$

    $-3x + 15y - 6z = \ 9$

11. _____

Chapter 3 SYSTEMS OF LINEAR EQUATIONS

3.3   Applications of Systems of Linear Equations

Learning Objectives

1	Solve geometry problems using two variables.
2	Solve money problems using two variables.
3	Solve mixture problems using two variables
4	Solve distance-rate-time problems using two variables.
5	Solve problems with three variables using a system of three equations.

Key Terms

Use the vocabulary terms listed below to complete each statement in exercises 1−2.

elimination method   substitution

1. Using the addition property to solve a system of equations is called the

   _____.

2. _____ is being used when one expression is replaced by another.

Objective 1   Solve geometry problems using two variables.

Video Examples

Review this example for Objective 1:

1. The length of a rectangular field is 5 m more than the width. Find the length and width if the perimeter is 70 m.

   Step 1 Read the problem. We must find the dimensions of the field.

   Step 2 Assign variables. Let L = length and W = width.

   Step 3 Write a system of equations. We use the equation for perimeter.
   $$2L + 2W = 70$$
   A second equation uses the information given.
   $$L = W + 5$$
   The system of equations is
   $$2L + 2W = 70 \quad (1)$$
   $$L = W + 5 \qquad (2)$$

   Step 4 Solve the system. Since equation (2) is solved for L, we use substitution.

Now Try:

1. The length of a rectangle is 7 ft more than the width. The perimeter is 54 ft. Find the dimensions of the rectangle.

   _____

$$2L + 2W = 70$$
$$2(W + 5) + 2W = 70$$
$$2W + 10 + 2W = 70$$
$$4W + 10 = 70$$
$$4W = 60$$
$$W = 15$$

Let $W = 15$ in equation (2) to find L.
$$L = 15 + 5 = 20$$

Step 5 State the answer. The length is 20 m and the width is 15 m.

Step 6 Check.
$$2(20) + 2(15) = 70$$
$$20 = 15 + 5$$

The answer is correct.

Objective 1 Practice Exercises

For extra help, see Example 1 on pages 240–241 of your text.

Solve each problem.

1. The side of a square is 5 centimeters shorter than the side of an equilateral triangle. The perimeter of the square is 7 centimeters less than the perimeter of the triangle. Find the lengths of a side of the square and of a side of the triangle.

   1. square: _____

      triangle: _____

2. The perimeter of a rectangle is 96 inches. If the width were tripled, the width would be 36 inches more than the length. Find the length and width of the rectangle.

   2. length: _____

      width: _____

3. The perimeter of a triangle is 70 centimeters. Two sides of the triangle have the same length. The third side is 7 centimeters longer than either of the equal sides. Find the length of the equal sides of the triangle.

3. _____

Objective 2   Solve money problems using two variables.

Video Examples

Review this example for Objective 2:

2. The total receipts for a basketball game were $4690.50. There were 723 tickets sold, some for children and some for adults. If the adult tickets cost $9.50 and the children's tickets cost $4, how many of each type were there?

Step 1 Read the problem. There are two unknowns.

Step 2 Assign variables.
   Let a = the number of adult tickets sold.
   Let c = the number of child tickets sold.

Step 3 Write a system of equations. We write one equation using the total number of tickets.
   $a + c = 723$
We write another equation using the cost.
   $9.50a + 4c = 4690.50$
The system of equations is
   $a + c = 723$ (1)
   $9.50a + 4c = 4690.50$ (2)

Step 4 Solve the system. To eliminate c, multiply equation (1) by –4, and add.
$$-4a - 4c = -2892 \quad \text{Multiply (1) by } -4.$$
$$\underline{9.5a + 4c = 4690.5 \quad (2)}$$
$$5.5a \quad\quad = 1798.5 \quad \text{Add.}$$
$$a = 327 \quad\quad \text{Divide by 5.5.}$$

Now Try:

2. The Garden Center ordered 6 ounces of marigold seed and 8 ounces of carnation seed, paying $214.54. They later ordered another 12 ounces of marigold seed and 18 ounces of carnation seed, paying $464.28. Find the price per ounce for each type of seed.

marigold   _____

carnation   _____

To find the value of c, let $a=327$ in equation (1).

$$327 + c = 723$$

$$c = 396$$

Step 5 State the answer. The number of adult tickets sold is 327 and the number of child tickets sold is 396.

Step 6 Check.

$$327 + 396 = 723$$

$$9.50(327) + 4(396) = 4690.50$$

The answer is correct.

Objective 2 Practice Exercises

For extra help, see Example 2 on page 242 of your text.

Solve each problem.

4. Pablo has some \$10-bills and some \$20-bills. The total value of the money is \$650, with a total of 40 bills. How many of each are there?

    4. \$10-bills _____

    \$20-bills _____

5. Big Giant Super Market will sell 5 large jars and 2 small jars of their peanut butter for \$36. They will also sell 2 large jars and 5 small jars for \$27. What is the price of each jar?

    5. small _____

    large _____

6. A taxi charges a flat rate plus a certain charge per mile. A trip of 7 miles costs \$5.30, while a trip of 3 miles costs \$3.70. Find the flat rate and the charge per mile.

    6. flat rate _____

    per mile _____

Objective 3   Solve mixture problems using two variables.

Video Examples

Review this example for Objective 3:

3. A 75% solution will be mixed with a 55% solution to get 70 liters of 63% solution? How many liters of the 55% and 75% solutions should be used?

Step 1 Read the problem. There are two solution strengths. We are looking for an "in between" strength.

Step 2 Assign variables.
   Let x = the number of liters of 75% solution.
   Let y = the number of liters of 55% solution.

Step 3 Write a system of equations.
Write one equation using the total amount.
   $x + y = 70$

Write each percent as a decimal and multiply each solution by its concentration.
   $0.75x + 0.55y = 0.63(70)$

The system of equations is
   $x + y = 70$ (1)
   $0.75x + 0.55y = 44.1$ (2)

Step 4 Solve the system. Multiply equation (2) by 100. Multiply equation (1) by –55 to eliminate y.

$$
\begin{array}{ll}
-55x - 55y = -3850 & \text{Multiply (1) by } -55. \\
\underline{75x + 55y = 4410} & \text{Multiply (2) by 100.} \\
20x = 560 & \text{Add.} \\
 x = 28 & \text{Divide by 20.}
\end{array}
$$

Substitute the 28 for x in equation (1) to find the value of y.
   $28 + y = 70$
   $y = 42$

Step 5 State the answer. The desired mixture will contain 28 liters of 75% solution and 42 liters of 55% solution.

Step 6 Check.
Total amount: 28 + 42 = 70
Total concentration: 0.75(28) + 0.55(42) = 44.1
The answer is correct.

Now Try:

3. How many liters of water should be added to 25% antifreeze solution to get 30 liters of a 20% solution? How many liters of 25% solution are needed?

water   _____

25% solution _____

Objective 3 Practice Exercises

For extra help, see Example 3 on pages 243–244 of your text.

Solve each problem.

7.   Jorge wishes to make 150 pounds of coffee blend
      that can be sold for $8 per pound. The blend will be
      a mixture of coffee worth $6 per pound and coffee
      worth $12 per pound. How many pounds of each
      kind of coffee should be used in the mixture?

      7.

      $6 coffee_____

      $12 coffee_____

8.   Bags of coffee worth $90 a bag must be mixed with
      coffee worth $75 a bag to get 50 bags worth $87 a
      bag. How many bags of each are needed?

      8.

      $90 coffee_____

      $75 coffee_____

9.   A pharmacist wants to add water to a solution that
      contains 80% medicine. She wants to obtain 12 oz.
      of a solution that is 20% medicine. How much water
      and how much of the 80% solution should she use?

      9.

      water_____

      80% solution _____

Objective 4   Solve distance-rate-time problems using two variables.

Video Examples

Review this example for Objective 4:

4. A train travels 600 kilometers in the same time that a truck travels 520 kilometers. Find the speed of the train and the truck if the train's average speed is 8 kilometers per hour faster than the truck's.

Step 1 Read the problem. We need to find the rate of each vehicle.

Step 2 Assign variables.

Let x = the rate of the train.
Let y = the rate of the truck.

	d	r	t
Train	600	x	$\dfrac{600}{x}$
Truck	520	y	$\dfrac{520}{y}$

Step 3 Write a system of equations. From comparing the two speeds we have an equation.

$$x = y + 8$$

Since both vehicles travel for the same time, we have a second equation.

$$\frac{600}{x} = \frac{520}{y}$$

Multiplying both sides by xy, we have

$$600y = 520x.$$

The system of equations is

$$x = y + 8 \qquad (1)$$
$$600y = 520x \quad (2)$$

Step 4 Solve the system. We solve the system by substitution. Replace x with $y + 8$ in equation (2).

$$600y = 520(y + 8)$$
$$600y = 520y + 4160$$
$$80y = 4160$$
$$y = 52$$

Because $x = y + 8$,

$$x = 52 + 8 = 60.$$

Step 5 State the answer. The train's rate is 60 km per hr. The truck's rate is 52 km per hr.

Now Try:

4. Ashley walks 10 miles in the same time that Taylor walks 6 miles. If Ashley walks 1 mile per hour less than twice Taylor's rate, what is the rate at which each walks?

Ashley   _____

Taylor   _____

Step 6 Check.

Train: $\dfrac{600}{60} = 10 \text{ hr}$ Truck: $\dfrac{520}{52} = 10 \text{ hr}$

The rate of the train is 8 km more than the rate of the truck.

Objective 4 Practice Exercises

For extra help, see Examples 4–5 on pages 244–246 of your text.

Solve each problem.

10.   Two cars start together and travel in the same direction, one going twice as fast as the other. At the end of 3 hours, they are 96 miles apart. How fast is each traveling?

10.

slower_____

faster_____

11.   Travis and his sister Kate jog to school daily. Travis jogs at 9 miles per hour, and Kate jogs at 5 miles per hour. When Travis reaches school, Kate is $\dfrac{1}{2}$ mile from the school. How far do Travis and Kate live from their school? How long does it take Travis to jog to school?

11. distance_____

time_____

Objective 5   Solve problems with three variables using a system of three equations.

Video Examples

Review this example for Objective 5:

6. Lee has some $5, $10, and $20-bills. He has a total of 51 bills, worth $795. The number of $5-bills is 25 less than the number of $20-bills. Find the number of each type of bill he has.

Step 1: Read the problem again. There are three unknowns.

Step 2: Assign variables.
Let x = the number of $5 bills,
let y = the number of $10 bills,
let z = the number of $20 bills.

Step 3: Write a system of three equations.
There are a total of 51 bills, so
$$x + y + z = 51 \quad (1)$$
The bills amounted to $795, so
$$5x + 10y + 20z = 795 \quad (2)$$
The number of $5-bills is 25 less than the number of $20-bills, so $x = z - 25$ or
$$x - z = -25 \quad (3)$$
The system is
$$x + y + z = 51 \quad (1)$$
$$5x + 10y + 20z = 795 \quad (2)$$
$$x \qquad - z = -25 \quad (3)$$

Step 4: Solve the system.
Eliminate y.

$$-10x - 10y - 10z = -510 \quad \text{Multiply (1) by } -10$$
$$\underline{5x + 10y + 20z = 795 \quad (2)}$$
$$-5x + 10z = 285 \quad (4)$$

Solve the system consisting of equations (3) and (4).

$$5x - 5z = -125 \quad \text{Multiply (3) by 5}$$
$$\underline{-5x + 10z = 285 \quad (4)}$$
$$5z = 160$$
$$z = 32$$

Substitute 32 for z in equation (3) and solve for x.
$$x - z = -25 \quad (3)$$
$$x - 32 = -25 \quad \text{Let } z = 32.$$
$$x = 7$$

Now Try:

6. The manager of the Sweet Candy Shop wishes to mix candy worth $4 per pound, $6 per pound, and $10 per pound to get 100 pounds of a mixture worth $7.60 per pound. The amount of $10 candy must equal the total amounts of the $4 and the $6 candy. How many pounds of each must be used?

$4 candy _____

$6 candy _____

$10 candy _____

Substitute 7 for x and 32 for z in equation (1) and solve for y.

$$x + y + z = 51 \quad (1)$$

$$7 + y + 32 = 51 \quad \text{Let } x = 7, \ z = 32.$$

$$y + 39 = 51$$

$$y = 12$$

Step 5: State the answer. Lee has 7 $5-bills, 12 $10-bills, and 32 $20-bills.

Step 6: Check that the total value of the bills is $795 and that the number of $5-bills is 25 less than the number of $20-bills.

Objective 5 Practice Exercises

For extra help, see Examples 6–7 on pages 247–250 of your text.

Solve each problem involving three unknowns.

12. Julie has $80,000 to invest. She invests part at 5%, one fourth this amount at 6%, and the balance 7%. Her total annual income from interest is $4700. Find the amount invested at each rate.

12. 5% _____

    6% _____

    7% _____

13. A merchant wishes to mix gourmet coffee selling for $8 per pound, $10 per pound, and $15 per pound to get 50 pounds of a mixture that can be sold for $11.70 per pound. The amount of the $8 coffee must be 3 pounds more than the amount of the $10 coffee. Find the number of pounds of each that must be used.

13. $8/lb _____

    $10/lb _____

    $15/lb _____

14. A boy scout troop is selling popcorn. There are three different kinds of popcorn in three different arrangements. Arrangement I contains 1 bag of cheddar cheese popcorn, 2 bags of caramel popcorn, and 3 bags of microwave popcorn. Arrangement II contains 3 bags of cheddar cheese popcorn, 1 bag of caramel popcorn, and 2 bags of microwave popcorn. Arrangement III contains 2 bags of cheddar cheese popcorn, 3 bags of caramel popcorn, and 1 bag of microwave popcorn. Jim needs 28 bags of cheddar cheese popcorn, 22 bags of caramel popcorn, and 22 bags of microwave popcorn to give as stocking stuffers for Christmas. How many of each arrangement should he buy?

14. I _____

    II _____

    III _____

Chapter 4 EXPONENTS, POLYNOMIALS, AND POLYNOMIAL FUNCTIONS

4.1   Integer Exponents and Scientific Notation

Learning Objectives
1 Use the product rule for exponents.
2 Define 0 and negative exponents.
3 Use the quotient rule for exponents.
4 Use the power rules for exponents.
5 Simplify exponential expressions.
6 Use the rules for exponents with scientific notation.

Key Terms

Use the vocabulary terms listed below to complete each statement in exercises 1−5.

> **exponent**   **base**   **product rule for exponents**
>
> **power rule for exponents**   **scientific notation**   **quotient rule for exponents**

1. A number written as $a \times 10^n$, where $1 \le |a| < 10$ and n is an integer, is written in

   _____.

2. The statement "If m and n are any integers, then $a^m \cdot a^n = a^{m+n}$" is an example of the _____.

3. The statement "If m and n are any integers and $a \ne 0$, then $\dfrac{a^m}{a^n} = a^{m-n}$" is an

   example of the _____.

4. The statement "If m and n are any integers, then $\left(a^m\right)^n = a^{mn}$" is an example of the _____.

5. In the expression a^m, a is the _____ and m is the _____.

Objective 1   Use the product rule for exponents.

Video Examples

Review this example for Objective 1:

1. Apply the product rule for exponents, if possible, in this case.

   $(-9p^4 q)(4p^6 q^3)$

   $(-9p^4 q)(4p^6 q^3) = -9(4)p^4 p^6 q^1 q^3$

   $\qquad = -36p^{10} q^4$

Now Try:

1. Apply the product rule for exponents, if possible, in this case.

   $(4p^5 q^2)(-6pq^4)$

   _____

Objective 1 Practice Exercises

For extra help, see Example 1 on page 266 of your text.

Use the product rule to simplify each expression, if possible. Write each answer in exponential form.

1. $7^4 \cdot 7^3$

1. _____

2. $(-2c^7)(-4c^8)$

2. _____

3. $(3k^7)(-8k^2)(-2k^9)$

3. _____

Objective 2   Define 0 and negative exponents.

Video Examples

Review these examples for Objective 2:	**Now Try:**
2. Evaluate.	2. Evaluate.

a. 9^0

$9^0 = 1$

a. 25^0

_____

b. $(-9)^0$

$(-9)^0 = 1$

b. $(-25)^0$

_____

c. -9^0

$-9^0 = -(9^0) = -1$

c. -25^0

_____

d. $7^0 + 13^0$

$7^0 + 13^0 = 1 + 1 = 2$

d. $100^0 + 6^0$

_____

3. Write with only positive exponents.

$(8z)^{-4}$

$(8z)^{-4} = \dfrac{1}{(8z)^4}, \ z \neq 0$

3.   Write with only positive exponents.

$(6z)^{-2}$

_____

Objective 2 Practice Exercises

For extra help, see Examples 2–4 on pages 267–268 of your text.

Evaluate the expression. Assume that all variables represent nonzero real numbers.

4. $2^0 + 6^0$

4. _____

Evaluate or simplify each expression, and write it using only positive exponents. Assume that all variables represent nonzero real numbers.

5. $\dfrac{2}{r^{-7}}$ **5.** _____

6. $-2k^{-4}$ **6.** _____

Objective 3   Use the quotient rule for exponents.

Video Examples

Review this example for Objective 3:

5. Apply the quotient rule for exponents, if possible, and write the result using only positive exponents.

$$\frac{p^7}{p^3}$$

$$\frac{p^7}{p^3} = p^{7-3} = p^4, \; p \neq 0$$

Now Try:

5. Apply the quotient rule for exponents, if possible, and write the result using only positive exponents.

$$\frac{q^8}{q^5}$$

_____

Objective 3 Practice Exercises

For extra help, see Example 5 on page 269 of your text.

Use the quotient rule to simplify each expression, if possible, and write it using only positive exponents. Assume that all variables represent nonzero real numbers.

7. $\dfrac{4k^7}{m^5}$ **7.** _____

8. $\dfrac{3^{-1}}{3^4}$ **8.** _____

9. $\dfrac{p^6}{p^{-2}}$ **9.** _____

Objective 4   Use the power rules for exponents.

Video Examples

Review this example for Objective 4:

6. Simplify, using the power rules.

$$\left(-\frac{5n^4}{y}\right)^3$$

$$\left(-\frac{5n^4}{y}\right)^3 = \frac{\left(-5n^4\right)^3}{y^3} = \frac{(-5)^3 n^{4\cdot3}}{y^3}$$

$$= \frac{-125n^{12}}{y^3}, \ y \neq 0$$

Now Try:

6. Simplify, using the power rules.

$$\left(-\frac{3p^4}{q}\right)^3$$

_____

Objective 4 Practice Exercises

For extra help, see Examples 6–7 on pages 270–271 of your text.

Simplify each expression, and write it using only positive exponents. Assume that all variables represent nonzero real numbers.

10. $\left(-2w^3 z^7\right)^4$

10. _____

11. $\left(\frac{-2a}{b^2}\right)^7$

11. _____

12. $\left(\frac{2}{7}\right)^{-3}$

12. _____

Objective 5   Simplify exponential expressions.

Video Examples

Review this example for Objective 5:

8. Simplify the expression so that no negative exponents appear in the final result. Assume that all variables represent nonzero real numbers.

$$\left(\frac{5x^3}{y^2}\right)^2 \left(\frac{3x^4}{y^{-3}}\right)^{-1}$$

Now Try:

8. Simplify the expression so that no negative exponents appear in the final result. Assume that all variables represent nonzero real numbers.

$$\left(\frac{2x^3}{y}\right)^2 \left(\frac{3x^5}{y^{-4}}\right)^{-1}$$

_____

$$\left(\frac{5x^3}{y^2}\right)^2\left(\frac{3x^4}{y^{-3}}\right)^{-1} = \frac{5^2\left(x^3\right)^2}{\left(y^2\right)^2}\cdot\frac{y^{-3}}{3x^4}$$

$$= \frac{25x^6}{y^4}\cdot\frac{y^{-3}}{3x^4}$$

$$= \frac{25}{3}x^{6-4}y^{-4-3}$$

$$= \frac{25}{3}x^2y^{-7}$$

$$= \frac{25x^2}{3y^7}$$

Objective 5 Practice Exercises

For extra help, see Example 8 on page 272 of your text.

Simplify each expression, and write it using only positive exponents. Assume that all variables represent nonzero real numbers.

13. $\dfrac{c^{10}\left(c^2\right)^3}{\left(c^3\right)^3\left(c^2\right)^{-9}}$

13. _____

14. $\left(a^{-1}b^{-2}\right)^{-4}$

14. _____

15. $\left(\dfrac{k^3t^4}{k^2t^{-1}}\right)^{-4}$

15. _____

Objective 6   Use the rules for exponents with scientific notation.

Video Examples

Review these examples for Objective 6:

9. Write each number in scientific notation.

   a. 970,000

   Step 1 Place a caret to the right of the 9 to mark the new location of the decimal point.

   $9_{\wedge}70,000$

   Step 2 Count from the decimal point, which is understood to be after the last 0, to the caret.

Now Try:

9. Write each number in scientific notation.

   a. 3,946,000

   _____

9$_\wedge$70,000

count 5 places

Step 3 Since 9.7 is to be made greater, the exponent on 10 is positive.

$970,000 = 9.7 \times 10^5$

b. 0.000064

0.00006$_\wedge$4   count 5 places

Since 6.4 is to be made less, the exponent on 10 is negative.

$0.000064 = 6.4 \times 10^{-5}$

b. 0.00048

_____

10. Write each number in standard notation.

a. 2.06×10^4

2.0600$_\wedge$

Move the decimal point 4 places to the right. Attach 0s as necessary.

$2.06 \times 10^4 = 20,600$

b. 3.41×10^{-5}

0$_\wedge$00003.41

Move the decimal point 5 places to the left. Attach 0s as necessary.

$3.41 \times 10^{-5} = 0.0000341$

10. Write each number in standard notation.

a. 9.45×10^6

_____

b. 8.04×10^{-5}

_____

Objective 6 Practice Exercises

For extra help, see Examples 9–12 on pages 274–275 of your text.

Write in standard notation.

16. 9×10^7

16. _____

17. 4.2×10^{-5}

17. _____

Evaluate. Write answer in scientific notation and in standard form.

18. $\dfrac{0.0021 \times 4800}{1,600,000 \times 0.000007}$

18. _____

Chapter 4 EXPONENTS, POLYNOMIALS, AND POLYNOMIAL FUNCTIONS

4.2   Adding and Subtracting Polynomials

Learning Objectives
1   Know the basic definitions for polynomials.
2   Add and subtract polynomials.

Key Terms

Use the vocabulary terms listed below to complete each statement in exercises 1−12.

term	coefficient (numerical coefficient)	
algebraic expression	polynomial	polynomial in x
descending powers	trinomial	binomial
monomial	degree of a term	degree of a polynomial
negative of a polynomial		

1.   The _____ is the sum of the exponents on the variables in that term.

2.   A polynomial in x is written in _____ if the exponents on x decrease from left to right.

3.   A(n) _____ is a number, a variable, or a product or quotient of a number and one or more variables raised to powers.

4.   A polynomial with exactly three terms is called a _____.

5.   A(n) _____ is a term, or a finite sum of terms, in which all variables have whole number exponents and no variables appear in denominators.

6.   The numerical factor in a term is its _____.

7.   A polynomial with exactly one term is called a _____.

8.   The _____ is the greatest degree of any term of the polynomial.

9.   A(n) _____ is a polynomial with exactly two terms.

10.   A(n) _____ is any combination of variables or constants joined by the basic operations of addition, subtraction, multiplication, and division (except by 0), or raising to powers or taking roots.

11. The _____ is obtained by changing the sign of every coefficient in the polynomial.

12. A polynomial containing only the variable x is a _____.

Objective 1   Know the basic definitions for polynomials.

Video Examples

Review these examples for Objective 1:

1. Write the polynomial in descending powers of the variable. Then give the leading term and the leading coefficient.

$$y^2 - 19y^4 + 7y^6 - 15y^3 + 17$$

Rewriting the expression, we have
$$7y^6 - 19y^4 - 15y^3 + y^2 + 17$$

The leading term is $7y^6$ and the leading coefficient is 7.

2. Identify the polynomial as a *monomial*, a *binomial*, a *trinomial*, or *none of these*. Also, give the degree.

$$-5x^3 + 7x + 4$$

This is a trinomial of degree 3.

Now Try:

1. Write the polynomial in descending powers of the variable. Then give the leading term and the leading coefficient.
$$x^3 - 5x^5 + 11x^7 - 4$$

_____

_____

_____

2. Identify the polynomial as a *monomial*, a *binomial*, a *trinomial*, or *none of these*. Also, give the degree.
$$-y^3 + 7y^2 + 4$$

_____

Objective 1 Practice Exercises

For extra help, see Examples 1–2 on page 282 of your text.

Write the polynomial in descending powers. Give the leading term and coefficient.

1. $8 + 5y - 7y^2 + y^3$ **1.** _____

Identify each polynomial as a **monomial***, a* **binomial***, a* **trinomial***, or* **none of these***. Give the degree.*

2. $\frac{1}{2}x^2 - \frac{3}{4}x + \frac{1}{4}x$ **2.** _____

3. $p + 3p^4$ **3.** _____

Objective 2   Add and subtract polynomials.

Video Examples

Review these examples for Objective 2:

3. Combine like terms.

  a. $8p + 7q - 10p + 3q$

$$8p + 7q - 10p + 3q = (8 - 10)p + (7 + 3)q$$
$$= -2p + 10q$$

  b. $12x^3y - 9xy^3 + 4x^3y - 5xy^3$

$$12x^3y - 9xy^3 + 4x^3y - 5xy^3$$
$$= 12x^3y + 4x^3y - 9xy^3 - 5xy^3$$
$$= 16x^3y - 14xy^3$$

4. Add $(7a^4 - 8a^2 + 17a) + (-18a^4 + 3a^2 + 6)$.

$$(7a^4 - 8a^2 + 17a) + (-18a^4 + 3a^2 + 6)$$
$$= 7a^4 - 18a^4 - 8a^2 + 3a^2 + 17a + 6$$
$$= -11a^4 - 5a^2 + 17a + 6$$

5. Subtract $(-7x^2 - 5x + 10) - (-8x^2 + 3x - 6)$.

$$(-7x^2 - 5x + 10) - (-8x^2 + 3x - 6)$$
$$= -7x^2 - 5x + 10 + 8x^2 - 3x + 6$$
$$= -7x^2 + 8x^2 - 5x - 3x + 10 + 6$$
$$= x^2 - 8x + 16$$

Now Try:

3. Combine like terms.

  a. $12m - 5n + 2m + 7n$

  _____

  b. $9x^2yz + 11xyz^2$
     $-10x^2yz + 2xyz^2$

  _____

4. Add $-7x^5 + 3x^3 - 4x$
     $\underline{9x^5 - 17x^3 + 3x}$

  _____

5. Subtract $(7y^3 - 4y^2 + 3y)$
     $-(-3y^3 + 8y^2 - 10y)$

  _____

Objective 2 Practice Exercises

For extra help, see Examples 3–5 on pages 283–284 of your text.

Add.

  4. $(x^2 + 6x - 8) + (3x^2 - 10)$

  4. _____

Name:

Date:

Instructor:

Section:

Subtract.

5.　$(5a^4 - 6a^2 + 9a) - (a^3 - 19a - 1)$

5. _____

Add or subtract as indicated.

6.　$4ab + 2bc - 9ac + 3ca - 2cb - 9ba$

6. _____

Chapter 4 EXPONENTS, POLYNOMIALS, AND POLYNOMIAL FUNCTIONS

4.3   Polynomial Functions, Graphs, and Composition

Learning Objectives	
1	Recognize and evaluate polynomial functions.
2	Use a polynomial function to model data.
3	Add and subtract polynomial functions.
4	Find the composition of functions.
5	Graph basic polynomial functions.

Key Terms

Use the vocabulary terms listed below to complete each statement in exercises 1–6.

polynomial function of degree *n*   **identity function**

squaring function   **cubing function**

composite function   **composition**

1.   The polynomial function defined by $f(x) = x^3$ is called the

_____.

2.   The function $g(f(x))$ is a _____.

3.   A function defined by $f(x) = a_n x^n + a_{n-1} x^{n-1} + \cdots + a_1 x + a_0$, where $a_n \neq 0$ and *n* is a whole number is a _____.

4.   The polynomial function defined by $f(x) = x^2$ is called the

_____.

5.   The simplest polynomial function is the _____ defined by $f(x) = x$.

6.   If *f* and *g* are functions, then the _____ of *g* and *f* is defined by $(g \circ f)(x) = g(f(x))$ for all *x* in the domain of *f* such that $f(x)$ is in the domain of *g*.

Objective 1   Recognize and evaluate polynomial functions.

Video Examples

Review this example for Objective 1:

1. Let $f(x) = 6x^3 - 6x + 1$. Find $f(-3)$.

Substitute -3 for x.

$$f(-3) = 6(-3)^3 - 6(-3) + 1$$
$$= 6(-27) - 6(-3) + 1$$
$$= -162 + 18 + 1$$
$$= -143$$

Now Try:

1. Let $p(x) = -x^4 + 3x^2 - x + 7$.
    Find $p(2)$.

    _____

Objective 1 Practice Exercises

For extra help, see Example 1 on page 287 of your text.

For each polynomial function, find (a) f(−2) and (b) f(3).

1. $f(x) = -x^2 - x - 5$

    1. (a) _____

        (b) _____

2. $f(x) = 2x^2 + 3x - 5$

    2. (a) _____

        (b) _____

3. $f(x) = 3x^4 - 5x^2$

    3. (a) _____

        (b) _____

Objective 2   Use a polynomial function to model data.

Video Examples

Review this example for Objective 2:

2. The average undergraduate tuition, room, and board at public four-year colleges for the years 1989–2009 can be modeled by the function

    $f(x) = 112.38x^2 - 2949.32x + 29,488.08$,

    where $x = 1$ corresponds to 1989, $x = 2$ corresponds to 1990, etc. Use this model to estimate the tuition, room, and board in 2004. (Source: National Center for Education Statistics)

    Since $x = 16$ corresponds to 2004, we must find $f(16)$.

    $f(16) = 112.38(16)^2 - 2949.32(16) + 29,488.08$
    $= 11,068.24$

    According to the model, the average cost of undergraduate tuition, room, and board at four-year public colleges was $11,068.24 in 2004.

Now Try:

2. Use the function at the left to estimate the tuition, room, and board in 2000.

    _____

Objective 2 Practice Exercises

For extra help, see Example 2 on page 287 of your text.

Solve each problem.

4. A widget manufacturer estimates that her monthly revenue can be modeled by the function

    $f(x) = -0.006x^2 + 32x - 10,000$.

    (a) Find the revenue if 2000 widgets are sold.

    (b) Find the revenue if 3500 widgets are sold.

4. a._____

    b._____

5. If a ball is batted at an angle of 35°, the distance that the ball travels is given approximately by

    $D = 0.029v^2 + 0.021v - 1$, where v is the bat speed in miles per hour and D is the distance traveled in feet. Find the distance a batted ball will travel if the ball is batted with a velocity of 90 miles per hour. Round your answer to the nearest whole number.

5. _____

6. The unemployment rate in a certain community can be modeled by the equation

$y = 0.0248x^2 - 0.4810x + 7.8543$, where y is the unemployment rate (percent) and x is the month ($x = 1$ represents January, $x = 2$ represents February, etc.) Use the model to find the unemployment rate in August. Round your answer to the nearest tenth.

6. _____

Objective 3   Add and subtract polynomial functions.

Video Examples

Review these examples for Objective 3:

3. For $f(x) = 2x^2 + 4x - 5$ and

$g(x) = -x^2 + 3x - 8$, find each of the following.

a. $(f + g)(x)$

$(f + g)(x) = f(x) + g(x)$

$\qquad = (2x^2 + 4x - 5) + (-x^2 + 3x - 8)$

$\qquad = x^2 + 7x - 13$

b. $(f - g)(x)$

$(f - g)(x) = f(x) - g(x)$

$\qquad = (2x^2 + 4x - 5) - (-x^2 + 3x - 8)$

$\qquad = (2x^2 + 4x - 5) + (x^2 - 3x + 8)$

$\qquad = 3x^2 + x + 3$

4. Find the following for polynomial functions f and g as defined.

$f(x) = 9x^2 - 3x$ and $g(x) = 5x.$

$(f + g)(4)$

$(f + g)(4) = f(4) + g(4)$

$\qquad = [9(4)^2 - 3(4)] + 5(4)$

$\qquad = [144 - 12] + 20$

$\qquad = 152$

Now Try:

3. For $f(x) = 6x^2 - 7x + 12$ and $g(x) = -3x^2 + x + 9$, find each of the following.

a. $(f + g)(x)$

_____

b. $(f - g)(x)$

_____

4. Find the following for polynomial functions f and g as defined.

$f(x) = 7x^2 - 5x$ and $g(x) = 4x.$

$(f + g)(x)$ and $(f + g)(-2)$

_____

Alternatively, we could first find $(f+g)(x)$.

$$(f+g)(x) = f(x) + g(x)$$
$$= (9x^2 - 3x) + 5x$$
$$= 9x^2 + 2x$$
$$(f+g)(4) = 9(4)^2 + 2(4)$$
$$= 152$$

Objective 3 Practice Exercises

For extra help, see Examples 3–4 on pages 288–289 of your text.

For the pair of functions, find (a) $(f+g)(x)$ and $(f-g)(x)$.

7. $f(x) = 2x^2 + 4x - 5,\ g(x) = -x^2 + 3x - 8$

7. a._____

   b._____

Let $f(x) = 3x^2 + 2$, $g(x) = -5x$, and $h(x) = x+2$. Find each of the following.

8. $(f-g)(-1)$

8. _____

9. $(f+h)(2)$

9. _____

Objective 4   Find the composition of functions.

Video Examples

Review these examples for Objective 4:

5. Let $f(x) = 3x - 1$ and $g(x) = x^2 + 2$. Find $(f \circ g)(5)$.

$$(f \circ g)(5) = f(g(5))$$
$$= f(5^2 + 2)$$
$$= f(27)$$
$$= 3(27) - 1$$
$$= 80$$

Now Try:

5. Let $f(x) = -3x - 3$ and $g(x) = x^2 - 5$. Find $(f \circ g)(-3)$.

   _____

7. Let $f(x) = 3x - 1$ and $g(x) = x^2 + 2$. Find the following.

$(g \circ f)(2)$

$(g \circ f)(2) = g(f(2))$

$\qquad = (f(2))^2 + 2$

$\qquad = (3(2) - 1)^2 + 2$

$\qquad = 5^2 + 2$

$\qquad = 27$

7. Let $f(x) = -3x - 3$ and $g(x) = x^2 - 5$. Find the following.

$(g \circ f)(2)$

_____

Objective 4 Practice Exercises

For extra help, see Examples 5–7 on pages 290–292 of your text.

Find the following.

10. Let $f(x) = 4x - 3$ and $g(x) = 2x^2 - 1$. Find the following.
    a. $(f \circ g)(2)$
    b. $(g \circ f)(-1)$
    c. $(f \circ g)(x)$

10. a._____

b._____

c._____

11. Let $f(x) = \dfrac{1}{x}$ and $g(x) = 3x^2 - 4x + 1$. Find the following.
    a. $(f \circ g)(2)$
    b. $(g \circ f)\left(\dfrac{1}{3}\right)$
    c. $(g \circ f)(x)$

11. a._____

b._____

c._____

Name: Date:
Instructor: Section:

Objective 5   Graph basic polynomial functions.

Video Examples

Review this example for Objective 5:

8. Graph $f(x) = x^2 + 2$. Give the domain and range.

For each input, square it and then add 2.

x	$f(x) = x^2 + 2$
-2	6
-1	3
0	2
1	3
2	6

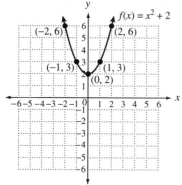

The domain is $(-\infty, \infty)$ and the range is $[2, \infty)$.

Now Try:

8. Graph $f(x) = -x^3 + 1$. Give the domain and range.

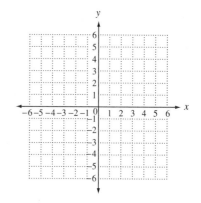

_____

Objective 5 Practice Exercises

For extra help, see Example 8 on page 293 of your text.

Graph each function by creating a table of ordered pairs. Give the domain and the range.

12. $f(x) = -3x + 2$

12. _____

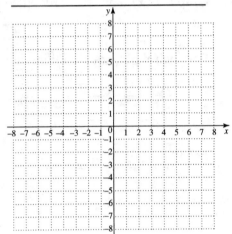

13. $f(x) = -2x^2$

13. _____

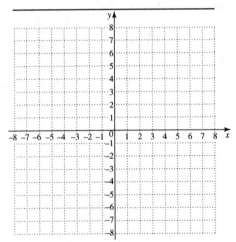

14. $f(x) = x^3 + 2$

14. _____

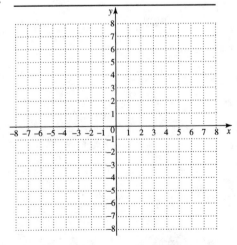

Chapter 4 EXPONENTS, POLYNOMIALS, AND POLYNOMIAL FUNCTIONS

4.4   Multiplying Polynomials

Learning Objectives
1   Multiply terms.
2   Multiply any two polynomials.
3   Multiply binomials.
4   Find the product of the sum and difference of two terms.
5   Find the square of a binomial.
6   Multiply polynomial functions.

Key Terms

Use the vocabulary terms listed below to complete each statement in exercises 1−3.

FOIL   **outer product**   **inner product**

1.   The _____ of $(2y-5)(y+8)$ is $-5y$.

2.   _____ is a shortcut method for finding the product of two binomials.

3.   The _____ of $(2y-5)(y+8)$ is $16y$.

Objective 1   Multiply terms.

Video Examples

Review this example for Objective 1:

1.   Find the product.

$$3p^3q^4(4p^4q^5)$$

$$3p^3q^4(4p^4q^5) = 3(4)p^3 \cdot p^4 \cdot q^4 \cdot q^5$$

$$= 12p^7q^9$$

Now Try:

1.   Find the product.

$$5x^4y^2(2x^5y)$$

_____

Objective 1 Practice Exercises

For extra help, see Example 1 on page 298 of your text.

Find each product.

1. $-2y^3(-8y^4)$

1.   _____

2. $3y^2z^3(6yz^4)$

2.   _____

3. $-12r^4s^7(-9r)$ **3.** _____

Objective 2   Multiply any two polynomials.

Video Examples

Review these examples for Objective 2:   **Now Try:**
2.   Find each product.   **2.**   Find each product.

a. $6x^3(-5x^3+7x-8)$ a. $7x^2(-4x^2-8x-6)$

$6x^3(-5x^3+7x-8)$
$\quad = 6x^3(-5x^3)+6x^3(7x)+6x^3(-8)$ _____
$\quad = -30x^6+42x^4-48x^3$

b. $5x^3(x+4)(x-7)$ b. $3x^2(x-1)(x+6)$

$5x^3(x+4)(x-7)$
$\quad = 5x^3\big[(x+4)(x)+(x+4)(-7)\big]$ _____
$\quad = 5x^3\big[x^2+4x-7x-28\big]$
$\quad = 5x^3(x^2-3x-28)$
$\quad = 5x^5-15x^4-140x^3$

Objective 2 Practice Exercises

For extra help, see Examples 2–3 on pages 298–299 of your text.

Find each product.

4. $7b^2(-5b^2+1-4b)$ **4.** _____

5. $(3m-5)(2m+4)$ **5.** _____

6. $3m^3+2m^2-4m$ **6.** _____
$\qquad\qquad 2m^2+1$

Objective 3 Multiply binomials.

Video Examples

Review this example for Objective 3:

4. Use the FOIL method to find the product.

$$(9x - 4y)(7x + 6y)$$

$(9x - 4y)(7x + 6y)$

    First   Outer   Inner   Last

$$= 63x^2 + 54xy - 28xy - 24y^2$$
$$= 63x^2 + 26xy - 24y^2$$

Now Try:

4. Use the FOIL method to find the product.

$$(6z - 4)(9z - 5)$$

_____

Objective 3 Practice Exercises

For extra help, see Example 4 on pages 300–301 of your text.

Find each product.

7. $(3x + 2y)(2x - 3y)$

7. _____

8. $(5a - b)(4a + 3b)$

8. _____

9. $(x - 5)(x + 3)$

9. _____

Objective 4 Find the product of the sum and difference of two terms.

Video Examples

Review these examples for Objective 4:

5. Find each product.

  a. $(8m + 5n)(8m - 5n)$

$$(8m + 5n)(8m - 5n) = (8m)^2 - (5n)^2$$
$$= 64m^2 - 25n^2$$

Now Try:

5. Find each product.

  a. $(3m + 8y)(3m - 8y)$

_____

b. $5x(x+4)(x-4)$

$$5x(x+4)(x-4) = 5x(x^2-16)$$
$$= 5x^3 - 80x$$

b. $10x^3(x+5)(x-5)$

_____

Objective 4 Practice Exercises

For extra help, see Example 5 on page 301 of your text.

Find each product.

10. $(8k+5p)(8k-5p)$

10. _____

11. $(7x-3y)(7x+3y)$

11. _____

12. $(9-4y)(9+4y)$

12. _____

Objective 5   Find the square of a binomial.

Video Examples

Review these examples for Objective 5:
6.   Find each product.

a. $(p-8)^2$

$$(p-8)^2 = p^2 - 2\cdot p\cdot 8 + 8^2$$
$$= p^2 - 16p + 64$$

b. $(5p+4q)^2$

$$(5p+4q)^2 = (5p)^2 + 2(5p)(4q) + (4q)^2$$
$$= 25p^2 + 40pq + 16q^2$$

Now Try:
6.   Find each product.

a. $(p-3)^2$

_____

b. $(4p-7q)^2$

_____

7. Use special products to find each product.

   a. $[(4p-5)+6q][(4p-5)-6q]$

   $[(4p-5)+6q][(4p-5)-6q]$

$$= (4p-5)^2 - (6q)^2$$
$$= 16p^2 - 40p + 25 - 36q^2$$

   b. $(a+3b)^4$

   $(a+3b)^4$

$$= (a+3b)^2 (a+3b)^2$$
$$= (a^2 + 6ab + 9b^2)(a^2 + 6ab + 9b^2)$$
$$= a^2(a^2 + 6ab + 9b^2) + 6ab(a^2 + 6ab + 9b^2)$$
$$+ 9b^2(a^2 + 6ab + 9b^2)$$
$$= a^4 + 6a^3b + 9a^2b^2 + 6a^3b + 36a^2b^2 + 54ab^3$$
$$+ 9a^2b^2 + 54ab^3 + 81b^4$$
$$= a^4 + 12a^3b + 54a^2b^2 + 108ab^3 + 81b^4$$

7. Use special products to find each product.

   a. $[(7p-3)+2q][(7p-3)-2q]$

   _____

   b. $(5a+b)^4$

   _____

Objective 5 Practice Exercises

For extra help, see Examples 6–7 on pages 302–303 of your text.

Find each square.

13. $(5y-3)^2$

13. _____

14. $(2p+3q)^2$

14. _____

15. $[(2x+3)-y]^2$

15. _____

Objective 6   Multiply polynomial functions.

Video Examples

Review this example for Objective 6:

8. For $f(x) = 5x + 2$ and $g(x) = 3x^2 + 4x$, find $(fg)(x)$ and $(fg)(-2)$.

$$(fg)(x) = f(x) \cdot g(x)$$
$$= (5x + 2)(3x^2 + 4x)$$
$$= 15x^3 + 20x^2 + 6x^2 + 8x$$
$$= 15x^3 + 26x^2 + 8x$$
$$(fg)(-2) = 15(-2)^3 + 26(-2)^2 + 8(-2)$$
$$= -120 + 104 - 16$$
$$= -32$$

Now Try:

8. For $f(x) = 6x + 5$ and $g(x) = 7x^2 + 2x$, find $(fg)(x)$ and $(fg)(-3)$.

_____

Objective 6 Practice Exercises

For extra help, see Example 8 on pages 303–304 of your text.

For the pair of functions, find the product $(fg)(x)$.

16. $f(x) = x + 2,\ g(x) = 3x - 2$ 16. _____

Let $f(x) = 3x^2 + 2, g(x) = -5x,$ *and* $h(x) = x + 2$. *Find each of the following.*

17. $(fg)(-1)$ 17. _____

18. $(fh)(2)$ 18. _____

Chapter 4   EXPONENTS, POLYNOMIALS, AND POLYNOMIAL FUNCTIONS

4.5   Dividing Polynomials

Learning Objectives

1   Divide a polynomial by a monomial.
2   Divide a polynomial by a polynomial of two or more terms.
3   Divide polynomial functions.

Key Terms

Use the vocabulary terms listed below to complete each statement in exercises 1–3.

quotient   **dividend**   **divisor**

1.   In the division $\dfrac{5x^5 - 10x^3}{5x^2} = x^3 - 2x$, the expression $5x^5 - 10x^3$ is the

_____.

2.   In the division $\dfrac{5x^5 - 10x^3}{5x^2} = x^3 - 2x$, the expression $x^3 - 2x$ is the

_____.

3.   In the division $\dfrac{5x^5 - 10x^3}{5x^2} = x^3 - 2x$, the expression $5x^2$ is the

_____.

Objective 1   Divide a polynomial by a monomial.

Video Examples

Review this example for Objective 1:

1.   Divide.

$$\frac{16x^2 - 8x + 24}{4}$$

$$\frac{16x^2 - 8x + 24}{4} = \frac{16x^2}{4} - \frac{8x}{4} + \frac{24}{4}$$

$$= 4x^2 - 2x + 6$$

Check: $4(4x^2 - 2x + 6) = 16x^2 - 8x + 24$

Now Try:

1.   Divide.

$$\frac{18x^2 + 9x - 27}{9}$$

_____

Name: Date:

Instructor: Section:

Objective 1 Practice Exercises

For extra help, see Example 1 on page 307 of your text.

Perform each division.

1. $\dfrac{12x^6 + 28x^5 + 20x^3}{4x^2}$

                                                    1. _____

2. $\dfrac{24w^8 + 12w^6 - 18w^4}{-6w^5}$

                                                    2. _____

3. $\dfrac{9r^2s + 18rs^2 - 27s^3}{-27rs^2}$

                                                    3. _____

Objective 2   Divide a polynomial by a polynomial of two or more terms.

Video Examples

Review these examples for Objective 2:

2.   Divide $\dfrac{3m^2 - 11m - 20}{m - 5}$.

$$
\begin{array}{r}
3m + 4 \\
m-5\overline{\smash{\big)}\,3m^2 - 11m - 20} \\
\underline{3m^2 - 15m} \\
4m - 20 \\
\underline{4m - 20} \\
0
\end{array}
$$

The quotient is $3m + 4$.

Now Try:

2.   Divide $\dfrac{5r^2 - 13r - 28}{r - 4}$.

                                     _____

3. Divide $4x^3 - 6x - 9$ by $x - 2$.

Add a term with 0 coefficient as a placeholder for the missing x^2-term.

$$
\begin{array}{r}
4x^2 + 8x + 10 \\
x - 2 \overline{\smash{\big)}\, 4x^3 + 0x^2 - 6x - 9} \\
\underline{4x^3 - 8x^2} \\
8x^2 - 6x \\
\underline{8x^2 - 16x} \\
10x - 9 \\
\underline{10x - 20} \\
11
\end{array}
$$

The quotient is $4x^2 + 8x + 10 + \dfrac{11}{x - 2}$.

4. Divide $8r^4 + 10r^3 - 6r^2 - 13r - 3$ by $2r^2 - 3$.

$$
\begin{array}{r}
4r^2 + 5r + 3 \\
2r^2 + 0r - 3 \overline{\smash{\big)}\, 8r^4 + 10r^3 - 6r^2 - 13r - 3} \\
\underline{8r^4 + 0r^3 - 12r^2} \\
10r^3 + 6r^2 - 13r \\
\underline{10r^3 + 0r^2 - 15r} \\
6r^2 + 2r - 3 \\
\underline{6r^2 + 0r - 9} \\
2r + 6
\end{array}
$$

The quotient is $4r^2 + 5r + 3 + \dfrac{2r + 6}{2r^2 - 3}$.

5. Divide $4p^3 - 11p^2 + 14p - 16$ by $4p - 8$.

$$
\begin{array}{r}
p^2 - \frac{3}{4}p + 2 \\
4p - 8 \overline{\smash{\big)}\, 4p^3 - 11p^2 + 14p - 16} \\
\underline{4p^3 - 8p^2} \\
-3p^2 + 14p \\
\underline{-3p^2 + 6p} \\
8p - 16 \\
\underline{8p - 16} \\
0
\end{array}
$$

The quotient is $p^2 - \dfrac{3}{4}p + 2$.

3. Divide $2x^3 - 15x - 7$ by $x - 3$.

_____

4. Divide
$12r^4 - 6r^3 - 16r^2 - 20r - 32$ by $2r^2 - 5$.

_____

5. Divide $5p^3 - 13p^2 + 26p - 40$ by $5p - 10$.

_____

Objective 2 Practice Exercises

For extra help, see Examples 2–5 on pages 308–310 of your text.

Perform each division.

4. $\dfrac{81a^2 - 1}{9a + 1}$ 4. _____

5. $\left(27p^4 - 36p^3 - 6p^2 + 23p - 20\right) \div \left(3p - 4\right)$ 5. _____

6. $\dfrac{6x^4 - 12x^3 + 13x^2 - 5x - 1}{2x^2 + 3}$ 6. _____

Name: _____   Date: _____

Instructor: _____   Section: _____

Objective 3   Divide polynomial functions.

Video Examples

Review this example for Objective 3:

6. For $f(x) = 4x^2 - 17x - 15$ and $g(x) = x - 5$, find $\left(\dfrac{f}{g}\right)(x)$ and $\left(\dfrac{f}{g}\right)(-2)$.

$$\left(\frac{f}{g}\right)(x) = \frac{f(x)}{g(x)} = \frac{4x^2 - 17x - 15}{x - 5}$$

$$
\begin{array}{r}
4x + 3 \\
x - 5 \overline{\smash{)}\, 4x^2 - 17x - 15} \\
\underline{4x^2 - 20x} \\
3x - 15 \\
\underline{3x - 15} \\
0
\end{array}
$$

$$\left(\frac{f}{g}\right)(x) = 4x + 3, \quad x \neq 5$$

$$\left(\frac{f}{g}\right)(-2) = 4(-2) + 3 = -5$$

Now Try:

6. For $f(x) = 2x^2 - 3x - 20$ and $g(x) = x - 4$, find $\left(\dfrac{f}{g}\right)(x)$ and $\left(\dfrac{f}{g}\right)(-2)$.

_____

Objective 3 Practice Exercises

For extra help, see Example 6 on pages 310–311 of your text.

For the pair of functions, find the quotient $\left(\dfrac{f}{g}\right)(x)$ and give any x-values that are not in the domain of the quotient function.

7. $f(x) = 4x^2 - 11x - 45,\ g(x) = x - 5$

7. _____

Let $f(x) = 4x^2 - 81, g(x) = 3x,$ and $h(x) = 2x + 9.$ *Find each of the following.*

8. $\left(\dfrac{f}{h}\right)(3)$

8. _____

9. $\left(\dfrac{g}{h}\right)(-2)$

9. _____

Chapter 5 FACTORING

5.1   Greatest Common Factors and Factoring by Grouping

Learning Objectives
1   Factor out the greatest common factor.
2   Factor by grouping.

Key Terms

Use the vocabulary terms listed below to complete each statement in exercises 1−3.

> **greatest common factor**   **factor**   **factored form**

1. An expression is in _____ when it is written as a product.

2. The _____ is the largest quantity that is a factor of each of a group of quantities.

3. An expression A is a _____ of an expression B if B can be divided by A with 0 remainder.

Objective 1   Factor out the greatest common factor.

Video Examples

Review these examples for Objective 1:

1. Factor out the greatest common factor.

   $7z - 35$

   GCF = 7
   $$7z - 35 = 7 \cdot z - 7 \cdot 5$$
   $$= 7(z - 5)$$

2. Factor out the greatest common factor.

   $24x^2 + 36x^3$

   The GCF is $12x^2$.
   $$24x^2 + 36x^3 = 12x^2(2) + 12x^2(3x)$$
   $$= 12x^2(2 + 3x)$$

Now Try:

1. Factor out the greatest common factor.
   $6z - 54$

   _____

2. Factor out the greatest common factor.
   $11x^2 + 44x$

   _____

3. Factor out the greatest common factor.

$$(x+4)(x+3)+(x+4)(3x+8)$$

The greatest common factor is $x+4$.
$$(x+4)(x+3)+(x+4)(3x+8)$$
$$=(x+4)[(x+3)+(3x+8)]$$
$$=(x+4)(4x+11)$$

4. Factor $-c^4+5c^3-9c^2$ two ways.

First, c^2 could be used as the common factor.
$$-c^4+5c^3-9c^2$$
$$=c^2(-c^2)+c^2(5c)+c^2(-9)$$
$$=c^2(-c^2+5c-9)$$

Because of the leading negative sign, $-c^2$ could be used as the common factor.
$$-c^4+5c^3-9c^2$$
$$=-c^2(c^2)-c^2(-5c)-c^2(9)$$
$$=-c^2(c^2-5c+9)$$

3. Factor out the greatest common factor.
$$(y+5)(y-9)+(y+5)(4y+5)$$

_____

4. Factor $-7b^4-8b^3+5b^2$ two ways.

_____

Objective 1 Practice Exercises

For extra help, see Examples 1–4 on pages 324–326 of your text.

Factor out the greatest common factor, if possible.

1. $2x^2y^8+5p^3q$

1. _____

2. $45a^2b^3-90ab+15ab^2$

2. _____

3. $(x+2)(2x+3)-(x+2)(x+1)$

3. _____

Objective 2   **Factor by grouping.**

Video Examples

Review these examples for Objective 2:

5. Factor $7x - 7y + ax - ay$.

Group the terms in pairs so that each pair has a common factor.
$7x - 7y + ax - ay$
$$= (7x - 7y) + (ax - ay)$$
$$= 7(x - y) + a(x - y)$$
$$= (x - y)(7 + a)$$

Check
$$(x - y)(7 + a) = 7x + ax - 7y - ay$$
$$= 7x - 7y + ax - ay$$

6. Factor $8m - 8n - pm + pn$.

$8m - 8n - pm + pn$
$$= (8m - 8n) + (-pm + pn)$$
$$= 8(m - n) - p(m - n)$$
$$= (m - n)(8 - p)$$
Check by multiplying.

7. Factor $5ax + 15ay + x + 3y$.

$5ax + 15ay + x + 3y$
$$= (5ax + 15ay) + (x + 3y)$$
$$= 5a(x + 3y) + 1(x + 3y)$$
$$= (x + 3y)(5a + 1)$$

8. Factor $x^2 y^2 - 12 + 4x^2 - 3y^2$.

$x^2 y^2 - 12 + 4x^2 - 3y^2$
$$= (x^2 y^2 - 3y^2) + (4x^2 - 12)$$
$$= y^2 (x^2 - 3) + 4(x^2 - 3)$$
$$= (x^2 - 3)(y^2 + 4)$$

Now Try:

5. Factor $9a - 9b + ra - rb$.

_____

6. Factor $ay - 5y - 3a + 15$.

_____

7. Factor $9px + 36qx + p + 4q$.

_____

8. Factor $a^2 b^2 - 48 - 6b^2 + 8a^2$.

_____

Objective 2 Practice Exercises

For extra help, see Examples 5–9 on pages 326–328 of your text.

Factor by grouping.

4. $3x^3 + 3xy^2 + 4x^2y + 4y^3$ 4. _____

5. $x - 8y^2 + 2xy^2 - 4$ 5. _____

6. $-3x - 6 + 2y + xy$ 6. _____

Chapter 5 FACTORING

5.2 Factoring Trinomials

Learning Objectives
1   Factor trinomials when the coefficient of the second-degree term is 1.
2   Factor trinomials when the coefficient of the second-degree term is not 1.
3   Use an alternative method for factoring trinomials.
4   Factor by substitution.

Key Terms

Use the vocabulary terms listed below to complete each statement in exercises 1–2.

prime polynomial   **factoring**

1. The process of writing a polynomial as a product is called _____.

2. A polynomial that cannot be factored with integer coefficients is a

    _____.

Objective 1   Factor trinomials when the coefficient of the second-degree term is 1.

Video Examples

Review these examples for Objective 1:

1. Factor each trinomial.

    a. $y^2 - 3y - 10$

    Step 1 Find pairs of integers whose product is -10.

Step 1 Find pairs of integers whose product is -10.	*Step 2* Write sums of those pairs of integers.
$10(-1)$	$10 + (-1) = 9$
$-10(1)$	$-10 + 1 = -9$
$5(-2)$	$5 + (-2) = 3$
$-5(2)$	$-5 + 2 = -3$

The integers 2 and –5 have the necessary product and sum.

$y^2 - 3y - 10$ factors as $(y - 5)(y + 2)$.

b. $r^2 + 9r + 18$

Look for two integers with a product of 18 and a sum of 9. Only the pair 6 and 3 have a sum of 9.

$r^2 + 9r + 18$ factors as $(r + 6)(r + 3)$.

Now Try:

1. Factor each trinomial.

    a. $x^2 - 3x - 28$

    _____

    b. $p^2 + 10p + 21$

    _____

2. Factor $m^2 + 4m + 5$.

Look for factors of 5, 5 and 1, or –5 and –1.

Neither pair has a sum of 4, so $m^2 + 4m + 5$ cannot be factored with integer coefficients and is prime.

3. Factor $x^2 + 7ax - 18a^2$.

Look at this trinomial as a trinomial in the form $x^2 + bx + c$, where $b = 7a$ and $c = -18a^2$.

Step 1 Find pairs of integers whose product is $-18a^2$.	*Step* 2 Write sums of those pairs of integers.
$18a(-a)$	$18a + (-a) = 17a$
$-18a(a)$	$-18a + a = -17a$
$9a(-2a)$	$9a + (-2a) = 7a$
$-9a(2a)$	$-9a + 2a = -7a$
$6a(-3a)$	$6a + (-3a) = 3a$
$-6a(3a)$	$-6a + 3a = -3a$

The expressions $9a$ and $-2a$ have the necessary product and sum.

$x^2 + 7ax - 18a^2$ factors as $(x + 9a)(x - 2a)$.

4. Factor $18y^3 + 18y^2 - 360y$.

First factor out the GCF.

$18y^3 + 18y^2 - 360y$

$= 18y(y^2 + y - 20)$

To factor $y^2 + y - 20$, look for two integers whose product is –20 and whose sum is 1. The necessary integers are 5 and –4.

$= 18y(y + 5)(y - 4)$

2. Factor $x^2 + 10x + 13$.

_____

3. Factor $x^2 + 11ax - 26a^2$.

_____

4. Factor $15y^3 - 30y^2 - 120y$.

_____

Objective 1 Practice Exercises

For extra help, see Examples 1–4 on pages 331–332 of your text.

Factor completely. If a polynomial cannot be factored, write prime.

1. $x^2 + 11x + 18$

1. _____

2. $x^2 + 14x - 49$ **2.** _____

3. $r^2s^2 + 4rs - 21$ **3.** _____

Objective 2   Factor trinomials when the coefficient of the second-degree term is not 1.

Video Examples

Review this example for Objective 2:

5. Factor $15x^2 - x - 2$.

Since $a = 15$, $b = -1$, and $c = -2$, the product ac is -30. The two integers whose product is -30 and whose sum is $b = -1$, are -6 and 5.

$$15x^2 - x - 2$$
$$= 15x^2 + 5x - 6x - 2$$
$$= 5x(3x + 1) - 2(3x + 1)$$
$$= (3x + 1)(5x - 2)$$

Now Try:

5. Factor $18x^2 + 9x - 5$.

_____

Objective 2 Practice Exercises

For extra help, see Example 5 on page 333 of your text.

Factor completely. If a polynomial cannot be factored, write prime.

4. $3y^2 + 13y + 4$ **4.** _____

5. $20r^2 - 28r - 3$ **5.** _____

6. $20x^2 + 39x - 11$ **6.** _____

Objective 3 Use an alternative method for factoring trinomials.

Video Examples

Review these examples for Objective 3:

6. Factor the trinomial.

$5x^2 + 8x + 3$

Addition signs are used, since all the signs in the trinomial are addition. The first two expressions have a product of $5x^2$, so they must be $5x$ and x.

$(5x + \underline{})(x + \underline{})$

The product of the last two terms must be 3, which means the numbers must be 3 and 1.

$(5x + 1)(x + 3)$ gives the wrong middle term
$$15x + x = 16x$$
$(5x + 3)(x + 1)$ gives the correct middle term
$$3x + 5x = 8x$$

Therefore, $5x^2 + 8x + 3$ factors as
$(5x + 3)(x + 1)$.

7. Factor $24x^2 + 7xy - 6y^2$.

There is no common factor (except 1). Try 6 and 4 for 24 and 3 and –2, or –3 and 2 for 6.

$(6x - 2y)(4x + 3y)$ Wrong: common factor
$(6x - 3y)(4x + 2y)$ Wrong: common factor

Try 8 and 3 for 24 with 3 and –2, or –3 and 2.

$(8x - 2y)(3x + 3y)$ Wrong: common factor
$(8x + 3y)(3x - 2y)$ Wrong middle term
$$-16xy + 9xy = -7xy$$

The last result differs from the correct middle term only in sign, so interchange the signs of the second terms in the factors.

$24x^2 + 7xy - 6y^2$ factors as $(8x - 3y)(3x + 2y)$.

9. Factor $30y^3 + 42y^2 - 36y$.

$30y^3 + 42y^2 - 36y$

$= 6y(5y^2 + 7y - 6)$

$= 6y(5y - 3)(y + 2)$

Now Try:

6. Factor the trinomial.

$8x^2 - 6x - 5$

_____

7. Factor $15x^2 + xy - 6y^2$.

_____

9. Factor $36x^3 + 63x^2 - 135x$.

_____

8. Factor $-5x^2 + 22x + 15$.

Factor out -1 first.
$$\begin{aligned} -5x^2 + 22x + 15 &= -1(5x^2 - 22x - 15) \\ &= -1(5x + 3)(x - 5) \\ &= -(5x + 3)(x - 5) \end{aligned}$$

8. Factor $-7x^2 - 9x + 10$.

_____

Objective 3 Practice Exercises

For extra help, see Examples 6–9 on pages 333–335 of your text.

Factor completely. If a polynomial cannot be factored, write prime.

7. $6p^2 - p - 15$

7. _____

8. $6x^2 - 5xy - y^2$

8. _____

9. $2a^3b - 10a^2b^2 + 12ab^3$

9. _____

Objective 4   Factor by substitution.

Video Examples

Review these examples for Objective 4:

10. Factor $3(x + 6)^2 - 4(x + 6) - 15$.

We let a substitution variable t represent $x + 6$.
$$3(x + 6)^2 - 4(x + 6) - 15$$
$$\begin{aligned} &= 3t^2 - 4t - 15 \\ &= (3t + 5)(t - 3) \\ &= [3(x + 6) + 5][(x + 6) - 3] \\ &= (3x + 18 + 5)(x + 6 - 3) \\ &= (3x + 23)(x + 3) \end{aligned}$$

Now Try:

10. Factor $12(z + 4)^2 - 11(z + 4) - 15$.

_____

11. Factor $10y^4 + 29y^2 - 21$.

$10y^4 + 29y^2 - 21.$
$= 10(y^2)^2 + 29y^2 - 21$
$= 10t^2 + 29t - 21 \qquad \text{Let } t = y^2.$
$= (5t - 3)(2t + 7)$
$= (5y^2 - 3)(2y^2 + 7)$

11. Factor $8y^4 - 6y^2 - 5$.

_____

Objective 4 Practice Exercises

For extra help, see Examples 10–11 on page 336 of your text.

Factor.

10. $8(p + 5)^2 + 2(p + 5) - 15$

10. _____

11. $8(5 - z)^2 - 14(5 - z) + 3$

11. _____

12. $4t^4 + 69t^2 + 17$

12. _____

Chapter 5 FACTORING

5.3   Special Factoring

Learning Objectives
1   Factor a difference of squares.
2   Factor a perfect square trinomial.
3   Factor a difference of cubes.
4   Factor a sum of cubes.

Key Terms

Use the vocabulary terms listed below to complete each statement in exercises 1−2.

perfect square trinomial   difference of squares

1. A _____ is a binomial that can be factored as the product of the sum and difference of two terms.

2. A _____ is a trinomial that can be factored as the square of a binomial.

Objective 1   Factor a difference of squares.

Video Examples

Review these examples for Objective 1:
1. Factor each polynomial.

    a. $t^2 - 25$

$$t^2 - 25 = t^2 - 5^2$$
$$= (t+5)(t-5)$$

    b. $9m^2 - 121p^2$

$$9m^2 - 121p^2 = (3m)^2 - (11p)^2$$
$$= (3m+11p)(3m-11p)$$

Now Try:
1. Factor each polynomial.

    a. $t^2 - 169$

    _____

    b. $25p^2 - 64q^2$

    _____

Objective 1 Practice Exercises

For extra help, see Example 1 on pages 338–339 of your text.

Factor each binomial completely. If a binomial cannot be factored, write prime.

1. $25a^2 - 36$

1. _____

2. $16y^4 - 81$ **2.** _____

3. $q^2 - (2r+3)^2$ **3.** _____

Objective 2   Factor a perfect square trinomial.

Video Examples

Review these examples for Objective 2:   **Now Try:**

2. Factor each polynomial.   **2.** Factor each polynomial.

    a. $16m^2 + 70m + 81$ **a.** $49a^2 + 56ab + 64b^2$

If this is a perfect square trinomial, it will equal
$(8m+9)^2$. The middle term is   _____
   $2(8m)(9) = 144m$, which does not equal $70m$.
The polynomial is prime.

    b. $x^2 - 10x + 25 - y^2$ **b.** $m^2 - 18m + 81 - b^2$

Since there are four terms, we use factoring by
grouping.   _____
$x^2 - 10x + 25 - y^2$
$= (x^2 - 10x + 25) - y^2$
$= (x-5)^2 - y^2$
$= (x-5+y)(x-5-y)$

Objective 2 Practice Exercises

For extra help, see Example 2 on pages 339–340 of your text.

Factor each polynomial completely.

4. $16q^2 - 40q + 25$ **4.** _____

5. $64p^4 + 48p^2q^2 + 9q^4$ **5.** _____

6. $(m-n)^2 - 12(m-n) + 36$ **6.** _____

Objective 3   Factor a difference of cubes.

Video Examples

Review these examples for Objective 3:

3.   Factor the polynomial.

$a^3 - 64$

$a^3 - 64 = a^3 - 4^3$

$\qquad = (a-4)(a^2 + 4a + 16)$

Now Try:

3.   Factor the polynomial.

$1000x^3 - y^3$

_____

Objective 3 Practice Exercises

For extra help, see Example 3 on pages 340–341 of your text.

Factor.

7. $8r^3 - 27s^3$ **7.** _____

8. $216m^3 - 125p^6$ **8.** _____

9. $8a^3 - 125b^3$ **9.** _____

Objective 4   Factor a sum of cubes.

Video Examples

Review these examples for Objective 4:

4.   Factor each polynomial.

   a. $r^3 + 64$

   $r^3 + 64 = r^3 + 4^3$

   $\qquad = (r+4)(r^2 - 4r + 16)$

   b. $(x+5)^3 + k^3$

   $(x+5)^3 + k^3$

   $\quad = [(x+5)+k][(x+5)^2 - (x+5)k + k^2]$

   $\quad = (x+5+k)(x^2 + 10x + 25 - xk - 5k + k^2)$

Now Try:

4.   Factor each polynomial.

   a. $r^3 + 125$

   _____

   b. $(a-5)^3 + b^3$

   _____

Objective 4 Practice Exercises

For extra help, see Example 4 on pages 341–342 of your text.

Factor.

10. $8a^3 + 64b^3$

10. _____

11. $125p^3 + q^3$

11. _____

12. $64x^3 + 343y^3$

12. _____

Chapter 5 FACTORING

5.4   A General Approach to Factoring

Learning Objectives
1   Factor out any polynomial.

Key Terms

Use the vocabulary terms listed below to complete each statement in exercises 1–2.

FOIL   factoring by grouping

1.   When there are more than three terms in a polynomial, use a process called
_____ to factor the polynomial.

2.   _____ is a shortcut method for finding the product of two
binomials.

Objective 1   Factor any polynomial.

Video Examples

Review these examples for Objective 1:

1. Factor each polynomial.

   a. $6x + 48$

   $6x + 48 = 6(x + 8)$

   $\qquad$ GCF $= 6$

   b. $15n^3 p^3 + 5n^2 p$

   $15n^3 p^3 + 5n^2 p = 5n^2 p(3np^2 + 1)$

   $\qquad\qquad$ GCF $= 5n^2 p$

   c. $9x(a + c) - z(a + c)$

   Factor out $(a + c)$
   $9x(a + c) - z(a + c) = (a + c)(9x - z)$

2. Factor each binomial if possible.

   a. $81m^2 - 49n^2$

   Difference of squares
   $81m^2 - 49n^2 = (9m)^2 - (7n)^2$
   $\qquad\qquad = (9m + 7n)(9m - 7n)$

Now Try:

1. Factor each polynomial.

   a. $18x + 54$

   _____

   b. $18r^4 s^2 + 6r^3 s$

   _____

   c. $7x(y + z) - 5(y + z)$

   _____

2. Factor each binomial if possible.

   a. $121x^2 - 25y^2$

   _____

b. $216p^3 - 125z^3$

Difference of cubes

$216p^3 - 125z^3$

$= (6p)^3 - (5z)^3$

$= (6p - 5z)\left[(6p)^2 + (6p)(5z) + (5z)^2\right]$

$= (6p - 5z)(36p^2 + 30pz + 25z^2)$

c. $16y^2 + 49$

$16y^2 + 49$ is prime. It is the sum of squares. There is no common factor.

3. Factor each trinomial.

a. $64z^2 - 80z + 25$

Perfect square trinomial

$64z^2 - 80z + 25 = (8z - 5)^2$

b. $36y^2 + 33yz - 15z^2$

$36z^2 + 33z - 15 = 3(12z^2 + 11z - 5)$

$\qquad\qquad\quad = 3(4z + 5)(3z - 1)$

c. $y^2 - 3y - 4$

$y^2 - 3y - 4 = (y - 4)(y + 1)$

4. Factor each polynomial.

a. $4m^3 - 4m^2n + mn^2 - n^3$

Group the terms and factor each group.

$4m^3 - 4m^2n + mn^2 - n^3$

$= (4m^3 - 4m^2n) + (mn^2 - n^3)$

$= 4m^2(m - n) + n(m - n)$

$= (m - n)(4m^2 + n^2)$

b. $27t^3 - 64w^3$

_____

c. $36p^2 + 169$

_____

3. Factor each trinomial.

a. $36z^2 - 84z + 49$

_____

b. $30y^2 - 5yz - 10z^2$

_____

c. $4k^2 - 7k - 2$

_____

4. Factor each polynomial.

a. $3c^3 - cd^2 + 3c^2d - d^3$

_____

b. $25b^2 + 10b + 1 - c^2$

Group the first three terms.

$25b^2 + 10b + 1 - c^2$

$\quad = (25b^2 + 10b + 1) - c^2$

$\quad = (5b+1)^2 - c^2$

$\quad = (5b+1+c)(5b+1-c)$

c. $27x^3 - y^3 + 9x^2 - y^2$

Rearrange and group the terms.

$27x^3 - y^3 + 9x^2 - y^2$

$\quad = (27x^3 - y^3) + (9x^2 - y^2)$

$\quad = (3x-y)(9x^2 + 3xy + y^2) + (3x-y)(3x+y)$

$\quad = (3x-y)(9x^2 + 3xy + y^2 + 3x + y)$

b. $36b^2 - 12b + 1 - c^2$

_____

c. $125a^3 - b^3 + 25a^2 - b^2$

_____

Objective 1 Practice Exercises

For extra help, see Examples 1–4 on pages 345–347 of your text.

Factor completely.

1. $12a^2b^2 + 3a^2b - 9ab^2$

    1. _____

2. $2x^3y^4 - 72xy^2$

    2. _____

3. $128x^3 - 2y^3$

    3. _____

4. $2a^2 - 17a + 30$

    4. _____

5. $x^3 - 3x^2 + 7x - 21$

    5. _____

6. $a^2 - 6ab + 9b^2 - 25$

    6. _____

Chapter 5   FACTORING

5.5   Solving Equations by the Zero-Factor Property

Learning Objectives
1   Learn and use the zero-factor property.
2   Solve applied problems that require the zero-factor property.
3   Solve a formula for a specified variable, where factoring is necessary.

Key Terms

Use the vocabulary terms listed below to complete each statement in exercises 1–4.

quadratic equation   **standard form**   **double solution**

zero-factor property

1.   An equation written in the form $ax^2 + bx + c = 0$ is written in the
     _____ of a quadratic equation.

2.   An equation that can written in the form $ax^2 + bx + c = 0$, with $a \neq 0$, is a

     _____.

3.   The _____ states that if two number have a product
     of 0, then at least one of the numbers must be 0.

4.   When a quadratic equation has only one distinct solution, that number is a

     _____.

Objective 1   Learn and use the zero-factor property.

Video Examples

Review these examples for Objective 1:

1.   Solve $(x + 7)(3x - 5) = 0$.

     Use the zero-factor property.
     $$(x + 7)(3x - 5) = 0$$
     $$x + 7 = 0 \quad \text{or} \quad 3x - 5 = 0$$
     $$x = -7 \quad \text{or} \quad 3x = 5$$
     $$x = \frac{5}{3}$$

     Check $(x + 7)(3x - 5) = 0$

     $$(-7 + 7)[3(-7) - 5] \overset{?}{=} 0 \quad \left| \quad \left(\frac{5}{3} + 7\right)\left[3\left(\frac{5}{3}\right) - 5\right] \overset{?}{=} 0 \right.$$

     $$(0)(-26) \overset{?}{=} 0 \quad \left| \quad \frac{26}{3}(0) \overset{?}{=} 0 \right.$$

     $$\text{True} \quad 0 = 0 \quad \left| \quad \text{True} \quad 0 = 0 \right.$$

     The solution set is $\left\{-7, \frac{5}{3}\right\}$.

Now Try:

1.   Solve $(4x + 5)(2x - 3) = 0$.

     _____

2. Solve each equation.

  a. Solve $3x^2 + 7x = 6$.

  Step 1 $\qquad\qquad 3x^2 + 7x = 6$

$\qquad\qquad\qquad\quad 3x^2 + 7x - 6 = 0$

  Step 2 $\qquad\quad (x+3)(3x-2) = 0$

  Step 3 $\quad x + 3 = 0 \quad$ or $\quad 3x - 2 = 0$

  Step 4 $\qquad x = -3 \quad$ or $\quad x = \dfrac{2}{3}$

Step 5 Check each solution in the original equation.

Check $\qquad\qquad 3x^2 + 7x = 6$

$3(-3)^2 + 7(-3) \overset{?}{=} 6 \quad\Big|\quad 3\left(\dfrac{2}{3}\right)^2 + 7\left(\dfrac{2}{3}\right) \overset{?}{=} 6$

$\qquad 27 - 21 \overset{?}{=} 6 \quad\Big|\qquad\quad \dfrac{4}{3} + \dfrac{14}{3} \overset{?}{=} 6$

$\qquad$ True $\quad 6 = 6 \quad\Big|\qquad$ True $\quad 6 = 6$

The solution set is $\left\{-3,\ \dfrac{2}{3}\right\}$.

  b. Solve $36x^2 = 60x - 25$.

$\qquad 36x^2 - 60x + 25 = 0$

$\qquad\qquad (6x - 5)^2 = 0$

$\qquad\qquad\quad 6x - 5 = 0$

$\qquad\qquad\qquad\quad 6x = 5$

$\qquad\qquad\qquad\quad x = \dfrac{5}{6}$

There is only one distinct solution, which we call a double solution. The solution set is $\left\{\dfrac{5}{6}\right\}$.

2. Solve each equation.

  a. Solve $4x^2 + 15x = 4$.

  _____

  b. Solve $49x^2 = 42x - 9$.

  _____

5. Solve $(3q+7)(q-1)=5(q+1)-2.$

$$(3q+7)(q-1)=5(q+1)-2$$

$$3q^2+4q-7=5q+5-2$$

$$3q^2+4q-7=5q+3$$

$$3q^2-q-10=0$$

$$(3q+5)(q-2)=0$$

$$3q+5=0 \quad \text{or} \quad q-2=0$$

$$q=-\frac{5}{3} \quad \text{or} \quad q=2$$

A check shows the solution set is $\left\{-\frac{5}{3},\ 2\right\}.$

5. Solve
$(3x+3)(2x-5)=5(1-x)-6x.$

_____

6. Solve $-x^3+x^2=-20x.$

$$-x^3+x^2+20x=0$$

$$x^3-x^2-20x=0$$

$$x(x^2-x-20)=0$$

$$x(x+4)(x-5)=0$$

$$x=0 \quad \text{or} \quad x+4=0 \quad \text{or} \quad x-5=0$$

$$x=-4 \qquad x=5$$

A check shows that the solution set is $\{-4, 0, 5\}.$

6. Solve $-x^3+2x^2=-35x.$

_____

Objective 1 Practice Exercises

For extra help, see Examples 1–6 on pages 349–352 of your text.

Solve each equation.

1. $2x^2-3x-20=0$

1. _____

2. $15x^2=x^3+56x$

2. _____

3. $z^2=6z-9$

3. _____

Objective 2   Solve applied problems that require the zero-factor property.

Video Examples

Review this example for Objective 2:

7. A house has a floor area of 608 square meters. The floor has the shape of a rectangle whose length is 13 meters more than the width. Find the width and length of the floor.

    Step 1 Read the problem again. There will be two answers.

    Step 2 Assign a variable. Let w = the width. Then $w + 13$ = the length.

    Step 3 Write an equation. The area is $A = lw$. Here, $608 = (w+13)w$

    Step 4 Solve.

    $$608 = w^2 + 13w$$
    $$0 = w^2 + 13w - 608$$
    $$0 = (w+32)(w-19)$$
    $$w + 32 = 0 \quad \text{or} \quad w - 19 = 0$$
    $$w = -32 \quad \text{or} \quad w = 19$$

    Step 5 State the answer. A distance cannot be negative, so reject –32 as a solution. The width is 19 m. The length is 19 + 13 = 32 m.

    Step 6 Check. $32(19) = 608 \text{ m}^2$

Now Try:

7. Paul and Joan wish to buy floor covering that covers 150 square feet for their large recreation room. They wish to cover a rectangle that is 5 feet longer than it is wide. How wide should the rectangle be?

    _____

Objective 2 Practice Exercises

For extra help, see Examples 7–8 on pages 353–354 of your text.

Solve each problem.

4. The area of a triangle is 42 square centimeters. The base is 2 centimeters less than twice the height. Find the base and height of the triangle.

    4. _____

5. The Browns installed 96 feet of fencing around a rectangular play yard. If the yard covers 540 square feet, what are its dimensions?

    5. _____

    _____

6. A company determines that its daily revenue R (in dollars) for selling x items is modeled by the equation $R = x(150 - x)$. How many items must be sold for its revenue to be $4400?

6. _____

Objective 3   Solve a formula for a specified variable, where factoring is necessary.

Video Examples

Review this example for Objective 3:

9. Solve the formula for H.
$$S = 2HW + 2LW + 2LH$$

$$S - 2LW = 2HW + 2LH$$

$$S - 2LW = H(2W + 2L)$$

$$\frac{S - 2LW}{2W + 2L} = H$$

Now Try:

9. The formula for the surface area of an open box is
$S = 2HW + LW + 2LH$, solve for W.

_____

Objective 3 Practice Exercises

For extra help, see Example 10 on page 354 of your text.

Solve each equation for the specified variable.

7. $bd = c + ba$, for b

7. _____

8. $n - mk = mt^2$, for m

8. _____

9. $v - w = uvw$, for w

9. _____

Chapter 6 RATIONAL EXPRESSIONS AND FUNCTIONS

6.1   Rational Expressions and Functions; Multiplying and Dividing

Learning Objectives
1   Define rational expressions.
2   Define rational functions and give their domains.
3   Write rational expressions in lowest terms.
4   Multiply rational expressions.
5   Find reciprocals of rational expressions.
6   Divide rational expressions.

Key Terms

Use the vocabulary terms listed below to complete each statement in exercises 1−2.

> **rational expression**   **rational function**

1.   A _____ is a function that is defined by

a rational expression in the form $f(x) = \dfrac{P(x)}{Q(x)}$, where $Q(x) \neq 0$.

2.   The quotient of two polynomials with denominator not 0 is called a

_____ .

Objective 1   Define rational expressions.

For extra help, see page 366 of your text.

Objective 2   Define rational functions and give their domains.

Video Examples

Review this example for Objective 2:

1.   For the rational function, find all numbers that are not in the domain. Then give the domain, using set-builder notation.

$$f(x) = \frac{5}{3x-9}$$

Set the denominator equal to 0 and solve.
$$3x - 9 = 0$$
$$3x = 9$$
$$x = 3$$

The number 3 cannot be used for a replacement for x. The domain of f includes all real numbers except 3, written using set-builder notation as $\{x \mid x \neq 3\}$.

Now Try:

1.   For the rational function, find all numbers that are not in the domain. Then give the domain, using set-builder notation.

$$f(x) = \frac{9}{5x-10}$$

_____

_____

Objective 2 Practice Exercises

For extra help, see Example 1 on pages 366–367 of your text.

Find all numbers that are not in the domain of each function. Then give the domain using set-builder notation.

1. $f(s) = \dfrac{8s + 7}{3s - 2}$ 1. _____

                                                   _____

2. $f(x) = \dfrac{x - 6}{x^2 + 1}$ 2. _____

                                                   _____

3. $f(q) = \dfrac{q + 7}{q^2 - 3q + 2}$ 3. _____

                                                     _____

Objective 3   **Write rational expressions in lowest terms.**

Video Examples

Review these examples for Objective 3:

2.   Write the rational expression in lowest terms.

$$\frac{(x+7)(x-1)}{(x-1)(x+6)}$$

$$\frac{(x+7)(x-1)}{(x-1)(x+6)} = \frac{(x+7)(x-1)}{(x+6)(x-1)} = \frac{x+7}{x+6}$$

3.   Write the rational expression in lowest terms.

$$\frac{p-25}{25-p}$$

$$\frac{p-25}{25-p} = \frac{p-25}{-1(p-25)} = \frac{1}{-1} = -1$$

Now Try:

2.   Write the rational expression in lowest terms.

$$\frac{(x-4)(x+3)}{(x+3)(x-7)}$$

                  _____

3.   Write the rational expression in lowest terms.

$$\frac{w-6}{6-w}$$

                  _____

Objective 3 Practice Exercises

For extra help, see Examples 2–3 on pages 368–369 of your text.

Write each rational expression in lowest terms.

4. $\dfrac{12k^3 + 12k^2}{3k^2 + 3k}$

4. _____

5. $\dfrac{2y^2 - 3y - 5}{2y^2 - 11y + 15}$

5. _____

6. $\dfrac{a^2 - 3a}{3a - a^2}$

6. _____

Objective 4   Multiply rational expressions.

Video Examples

Review this example for Objective 4:
4.   Multiply.

$$\frac{7r - 21}{r} \cdot \frac{3r^2}{8r - 24}$$

$$\frac{7r - 21}{r} \cdot \frac{3r^2}{8r - 24} = \frac{7(r - 3)}{r} \cdot \frac{3r \cdot r}{8(r - 3)}$$

$$= \frac{r(r - 3)}{r(r - 3)} \cdot \frac{7 \cdot 3r}{8}$$

$$= \frac{21r}{8}$$

Now Try:
4.   Multiply.

$$\frac{9x - 36}{x} \cdot \frac{4x^2}{7x - 28}$$

_____

Objective 4 Practice Exercises

For extra help, see Example 4 on page 371 of your text.

Multiply. Write each answer in lowest terms.

7. $\dfrac{x^2+x-12}{x^2+7x+10} \cdot \dfrac{x^2+3x-10}{x^2+2x-8}$

7. _____

8. $\dfrac{x^2+10x+21}{x^2+14x+49} \cdot \dfrac{x^2+12x+35}{x^2-6x-27}$

8. _____

9. $\dfrac{3m^2-m-10}{2m^2-7m-4} \cdot \dfrac{4m^2-1}{6m^2+7m-5}$

9. _____

Objective 5   Find reciprocals of rational expressions.

Objective 5 Practice Exercises

For extra help, see page 372 of your text.

Find the reciprocal.

10. $\dfrac{r^2+2r}{5+r}$

10. _____

11. $\dfrac{7z+7}{z^2-9}$

11. _____

12.   0

12. _____

Objective 6   Divide rational expressions.

Video Examples

Review this example for Objective 6:

5.   Divide.

$$\frac{8x^2}{21} \div \frac{2x}{7}$$

$$\frac{8x^2}{21} \div \frac{2x}{7} = \frac{8x^2}{21} \cdot \frac{7}{2x} = \frac{4x \cdot 2x}{3 \cdot 7} \cdot \frac{7}{2x} = \frac{4x}{3}$$

Now Try:

5.   Divide.

$$\frac{10p}{21} \div \frac{15p^2}{7}$$

_____

Objective 6 Practice Exercises

For extra help, see Example 5 on pages 372–373 of your text.

Divide. Write each answer in lowest terms.

13. $\dfrac{4m-12}{2m+10} \div \dfrac{9-m^2}{m^2-25}$

13. _____

14. $\dfrac{27-3k^2}{3k^2+8k-3} \div \dfrac{k^2-6k+9}{6k^2-19k+3}$

14. _____

15. $\dfrac{y^2+7y+10}{3y+6} \div \dfrac{y^2+2y-15}{4y-4}$

15. _____

197

Chapter 6 RATIONAL EXPRESSIONS AND FUNCTIONS

6.2   Adding and Subtracting Rational Expressions

Learning Objectives
1   Add and subtract rational expressions with the same denominator.
2   Find a least common denominator.
3   Add and subtract rational expressions with different denominators.

Key Terms

Use the vocabulary terms listed below to complete each statement in exercises 1−2.

least common denominator (LCD)   **equivalent expressions**

1. $\dfrac{24x-8}{9x^2-1}$ and $\dfrac{8}{3x+1}$ are _____.

2.   The simplest expression that is divisible by all denominators is called the

_____.

Objective 1   Add and subtract rational expressions with the same denominator.

Video Examples

Review these examples for Objective 1:
1.   Add or subtract as indicated.

a. $\dfrac{13}{3k^2} - \dfrac{19}{3k^2}$

$\dfrac{13}{3k^2} - \dfrac{19}{3k^2} = \dfrac{13-19}{3k^2} = \dfrac{-6}{3k^2}$, or $-\dfrac{2}{k^2}$

b. $\dfrac{a}{a^2-b^2} + \dfrac{b}{a^2-b^2}$

$\dfrac{a}{a^2-b^2} + \dfrac{b}{a^2-b^2} = \dfrac{a+b}{a^2-b^2}$

$= \dfrac{a+b}{(a+b)(a-b)}$

$= \dfrac{1}{a-b}$

Now Try:
1.   Add or subtract as indicated.

a. $\dfrac{16}{5z^2} - \dfrac{26}{5z^2}$

_____

b. $\dfrac{c}{c^2-d^2} - \dfrac{d}{c^2-d^2}$

_____

c. $\dfrac{5}{x^2+4x-5}+\dfrac{x}{x^2+4x-5}$

$$=\dfrac{5+x}{x^2+4x-5}$$

$$=\dfrac{5+x}{(x+5)(x-1)}$$

$$=\dfrac{1}{x-1}$$

c. $\dfrac{7}{x^2+6x-7}+\dfrac{x}{x^2+6x-7}$

_____

Objective 1 Practice Exercises

For extra help, see Example 1 on pages 376–377 of your text.

Add or subtract as indicated. Write each answer in lowest terms.

1. $\dfrac{n}{m+3}-\dfrac{-3n+7}{m+3}$

1. _____

2. $\dfrac{2x+3}{x^2+3x-10}+\dfrac{2-x}{x^2+3x-10}$

2. _____

3. $\dfrac{k}{k^2-6k+8}-\dfrac{2}{k^2-6k+8}$

3. _____

Objective 2   Find a least common denominator.

Video Examples

Review these examples for Objective 2:

2. Suppose that the given expressions are denominators of fractions. Find the LCD for each group of denominators.

  a. $3x^2y^3,\ 7x^3y^2$

$$3x^2y^3 = 3\cdot x^2\cdot y^3$$
$$7x^3y^2 = 7\cdot x^3\cdot y^2$$
$$\text{LCD} = 3\cdot 7\cdot x^3\cdot y^3 = 21x^3y^3$$

Now Try:

2. Suppose that the given expressions are denominators of fractions. Find the LCD for each group of denominators.

  a. $11x^2y,\ 3xy^4$

_____

b. $m-4,\ m$

Each denominator is already factored. The LCD must be divisible by both $m-4$ and m.

$m(m-4)$

c. $m^2+7m+12,\ m^2+6m+9,\ 5m+20$

$m^2+7m+12=(m+4)(m+3)$

$m^2+6m+9=(m+3)^2$

$5m+20=5(m+4)$

$\text{LCD}=5(m+4)(m+3)^2$

b. $y+6,\ y$

_____

c. $x^2-3x-4,\ x^2-8x+16,$

$3x+3$

_____

Objective 2 Practice Exercises

For extra help, see Example 2 on page 378 of your text.

Assume that the expressions given are denominators of fractions. Find the least common denominator (LCD) for each group.

4. $q^2-36,\ (q+6)^2$

4. _____

5. $x^2+5x+6,\ 3x+6$

5. _____

6. $p-4,\ p^2-16,\ (p+4)^2$

6. _____

Objective 3   Add and subtract rational expressions with different denominators.

Video Examples

Review these examples for Objective 3:

5. Add.

$$\frac{6p}{p-4}+\frac{2}{4-p} \quad \text{The LCD is } p-4.$$

Since the denominators are opposites, we multiply the second expression by -1.

$$=\frac{6p}{p-4}+\frac{2(-1)}{(4-p)(-1)}$$

$$=\frac{6p}{p-4}+\frac{-2}{p-4}$$

$$=\frac{6p-2}{p-4}$$

6. Add and subtract as indicated.

$$\frac{2}{x}-\frac{5}{x-2}+\frac{10}{x^2-2x} \quad \text{The LCD is } x(x-2).$$

$$=\frac{2}{x}-\frac{5}{x-2}+\frac{10}{x(x-2)}$$

$$=\frac{2(x-2)}{x(x-2)}-\frac{5x}{x(x-2)}+\frac{10}{x(x-2)}$$

$$=\frac{2x-4-5x+10}{x(x-2)}$$

$$=\frac{-3x+6}{x(x-2)}$$

$$=\frac{-3(x-2)}{x(x-2)}$$

$$=-\frac{3}{x}$$

7. Subtract.

$$\frac{7x}{x^2-4x+4}-\frac{2}{x^2-4}$$

$$=\frac{7x}{(x-2)^2}-\frac{2}{(x+2)(x-2)}$$

The LCD is $(x+2)(x-2)^2$.

$$=\frac{7x}{(x-2)^2}-\frac{2}{(x+2)(x-2)}$$

Now Try:

5. Add.

$$\frac{x}{x-8}+\frac{12}{8-x}$$

_____

6. Add and subtract as indicated.

$$\frac{12}{x^2+3x}-\frac{3}{x}+\frac{4}{x+3}$$

_____

7. Subtract.

$$\frac{8x}{x^2-10x+25}-\frac{3}{x^2-25}$$

_____

$$= \frac{7x(x+2)}{(x+2)(x-2)^2} - \frac{2(x-2)}{(x+2)(x-2)^2}$$

$$= \frac{7x(x+2) - 2(x-2)}{(x+2)(x-2)^2}$$

$$= \frac{7x^2 + 14x - 2x + 4}{(x+2)(x-2)^2}$$

$$= \frac{7x^2 + 12x + 4}{(x+2)(x-2)^2}$$

Objective 3 Practice Exercises

For extra help, see Examples 3–8 on pages 379–382 of your text.

Add or subtract as indicated. Write each answer in lowest terms.

7. $\dfrac{3}{n^2 - 16} - \dfrac{6n}{n^2 + 8n + 16}$

7. _____

8. $\dfrac{4z}{z^2 + 6z + 8} + \dfrac{2z - 1}{z^2 + 5z + 6}$

8. _____

9. $\dfrac{4y}{y^2 + 4y + 3} - \dfrac{3y + 1}{y^2 - y - 2}$

9. _____

Chapter 6 RATIONAL EXPRESSIONS AND FUNCTIONS

6.3   Complex Fractions

Learning Objectives
1   Simplify complex fractions by simplifying the numerator and denominator (Method 1).
2   Simplify complex fractions by multiplying by a common denominator (Method 2).
3   Compare the two methods of simplifying complex fractions.
4   Simplify rational expressions with negative exponents.

Key Terms

Use the vocabulary terms listed below to complete each statement in exercises 1−2.

complex fraction   LCD

1.   A _____ is a rational expression with one or more fractions in the numerator, denominator, or both.

2.   To simplify a complex fraction, multiply the numerator and denominator by the _____ of all the fractions within the complex fraction.

Objective 1   Simplify complex fractions by simplifying the numerator and denominator (Method 1).

Video Examples

Review these examples for Objective 1:

1.   Use Method 1 to simplify each complex fraction.

a. $\dfrac{\dfrac{x-2}{4x}}{\dfrac{x+3}{x}}$

$\dfrac{\dfrac{x-2}{4x}}{\dfrac{x+3}{x}} = \dfrac{x-2}{4x} \div \dfrac{x+3}{x}$

$= \dfrac{x-2}{4x} \cdot \dfrac{x}{x+3}$

$= \dfrac{x(x-2)}{4x(x+3)}$

$= \dfrac{x-2}{4(x+3)}$

Now Try:

1.   Use Method 1 to simplify each complex fraction.

a. $\dfrac{\dfrac{x+1}{3x}}{\dfrac{x-1}{5x}}$

_____

b. $\dfrac{\dfrac{4}{x}+3}{5-\dfrac{3}{x}}$

$$\dfrac{\dfrac{4}{x}+3}{5-\dfrac{3}{x}} = \dfrac{\dfrac{4}{x}+\dfrac{3x}{x}}{\dfrac{5x}{x}-\dfrac{3}{x}}$$

$$= \dfrac{\dfrac{4+3x}{x}}{\dfrac{5x-3}{x}}$$

$$= \dfrac{4+3x}{x}\cdot\dfrac{x}{5x-3}$$

$$= \dfrac{3x+4}{5x-3}$$

b. $\dfrac{2-\dfrac{7}{y}}{7-\dfrac{2}{y}}$

_____

Objective 1 Practice Exercises

For extra help, see Example 1 on pages 385–386 of your text.

Use Method 1 to simplify each complex fraction.

1. $\dfrac{\dfrac{3a+4}{a}}{\dfrac{1}{a}+\dfrac{2}{5}}$

1. _____

2. $\dfrac{\dfrac{2}{a+2}-4}{\dfrac{1}{a+2}-3}$

2. _____

3. $\dfrac{\dfrac{a+2}{a-2}}{\dfrac{1}{a^2-4}}$

3. _____

Objective 2   Simplify complex fractions by multiplying by a common denominator (Method 2).

Video Examples

Review these examples for Objective 2: | **Now Try:**

2.   Use Method 2 to simplify each complex fraction.

a. $\dfrac{\dfrac{4}{x}+3}{5-\dfrac{3}{x}}$

$$\frac{\dfrac{4}{x}+3}{5-\dfrac{3}{x}}=\frac{\left(\dfrac{4}{x}+3\right)\cdot x}{\left(5-\dfrac{3}{x}\right)\cdot x}$$

$$=\frac{\dfrac{4}{x}\cdot x+3\cdot x}{5\cdot x-\dfrac{3}{x}\cdot x}$$

$$=\frac{4+3x}{5x-3},\ \text{or}\ \frac{3x+4}{5x-3}$$

b. $\dfrac{\dfrac{4}{x-4}+3x}{5x-\dfrac{3}{x-4}}$

$$\frac{\dfrac{4}{x-4}+3x}{5x-\dfrac{3}{x-4}}=\frac{\left(\dfrac{4}{x-4}+3x\right)\cdot(x-4)}{\left(5x-\dfrac{3}{x-4}\right)\cdot(x-4)}$$

$$=\frac{\dfrac{4}{x-4}\cdot(x-4)+3x\cdot(x-4)}{5x\cdot(x-4)-\dfrac{3}{x-4}\cdot(x-4)}$$

$$=\frac{4-3x(x-4)}{5x(x-4)-3}$$

$$=\frac{4-3x^2+12x}{5x^2-20x-3},\ \text{or}\ \frac{-3x^2+12x+4}{5x^2-20x-3}$$

Now Try:

2.   Use Method 2 to simplify each complex fraction.

a. $\dfrac{2-\dfrac{7}{y}}{7-\dfrac{2}{y}}$

_____

b. $\dfrac{\dfrac{1}{a-2}+4a}{\dfrac{1}{a-2}+3a}$

_____

Objective 2 Practice Exercises

For extra help, see Example 2 on pages 386–387 of your text.

Use Method 2 to simplify each complex fraction.

4. $\dfrac{2x-y^2}{x+\dfrac{y^2}{x}}$

4. _____

5. $\dfrac{r+\dfrac{3}{r}}{\dfrac{5}{r}+rt}$ 5. _____

6. $\dfrac{\dfrac{x-2}{x+2}}{\dfrac{x}{x-2}}$ 6. _____

Objective 3   Compare the two methods of simplifying complex fractions.

Video Examples

Review these examples for Objective 3:

3. Use both Method 1 and Method 2 to simplify each complex fraction.

    a. $\dfrac{\dfrac{15}{10k+10}}{\dfrac{5}{3k+3}}$

    Using Method 1

    $\dfrac{\dfrac{15}{10k+10}}{\dfrac{5}{3k+3}} = \dfrac{\dfrac{15}{10(k+1)}}{\dfrac{5}{3(k+1)}} = \dfrac{15}{10(k+1)} \div \dfrac{5}{3(k+1)}$

    $= \dfrac{15}{10(k+1)} \cdot \dfrac{3(k+1)}{5}$

    $= \dfrac{9}{10}$

    Using Method 2

    $\dfrac{\dfrac{15}{10k+10}}{\dfrac{5}{3k+3}} = \dfrac{\dfrac{15}{10(k+1)}}{\dfrac{5}{3(k+1)}} = \dfrac{\dfrac{15}{10(k+1)} \cdot 30(k+1)}{\dfrac{5}{3(k+1)} \cdot 30(k+1)}$

    $= \dfrac{45}{50} = \dfrac{9}{10}$

Now Try:

3. Use both Method 1 and Method 2 to simplify each complex fraction.

    a. $\dfrac{\dfrac{1}{r}}{\dfrac{1+r}{1-r}}$

    _____

b. $\dfrac{\dfrac{1}{m} - \dfrac{1}{n^2}}{\dfrac{1}{m^2} + \dfrac{1}{n}}$

b. $\dfrac{\dfrac{1}{a^2} + \dfrac{1}{b}}{\dfrac{1}{b^2} - \dfrac{1}{a}}$

_____

Using Method 1

$$\dfrac{\dfrac{1}{m} - \dfrac{1}{n^2}}{\dfrac{1}{m^2} + \dfrac{1}{n}} = \dfrac{\dfrac{n^2}{mn^2} - \dfrac{m}{mn^2}}{\dfrac{n}{m^2 n} + \dfrac{m^2}{m^2 n}} = \dfrac{\dfrac{n^2 - m}{mn^2}}{\dfrac{n + m^2}{m^2 n}}$$

$$= \dfrac{n^2 - m}{mn^2} \div \dfrac{n + m^2}{m^2 n}$$

$$= \dfrac{n^2 - m}{mn^2} \cdot \dfrac{m^2 n}{n + m^2}$$

$$= \dfrac{m(n^2 - m)}{n(n + m^2)}$$

Using Method 2

$$\dfrac{\dfrac{1}{m} - \dfrac{1}{n^2}}{\dfrac{1}{m^2} + \dfrac{1}{n}} = \dfrac{\left(\dfrac{1}{m} - \dfrac{1}{n^2}\right) \cdot m^2 n^2}{\left(\dfrac{1}{m^2} + \dfrac{1}{n}\right) \cdot m^2 n^2}$$

$$= \dfrac{\left(\dfrac{1}{m}\right) m^2 n^2 - \left(\dfrac{1}{n^2}\right) m^2 n^2}{\left(\dfrac{1}{m^2}\right) m^2 n^2 + \left(\dfrac{1}{n}\right) m^2 n^2}$$

$$= \dfrac{mn^2 - m^2}{n^2 + m^2 n}, \text{ or } \dfrac{m(n^2 - m)}{n(n + m^2)}$$

Objective 3 Practice Exercises

For extra help, see Example 3 on pages 387–388 of your text.

Use either method to simplify each complex fraction.

7. $\dfrac{\dfrac{25k^2 - m^2}{4k}}{\dfrac{5k + m}{7k}}$

7. _____

8. $\dfrac{\dfrac{4}{p} - 2p}{\dfrac{3 - p^2}{6}}$ **8.** _____

9. $\dfrac{\dfrac{4}{x} - \dfrac{1}{2}}{\dfrac{5}{x} + \dfrac{1}{3}}$ **9.** _____

Objective 4   Simplify rational expressions with negative exponents.

Video Examples

Review this example for Objective 4:

4. Simplify, using only positive exponents in the answer.

$$\frac{2x^{-1} + y^2}{z^{-3}}$$

$$= \frac{\dfrac{2}{x} + y^2}{\dfrac{1}{z^3}} = \frac{xz^3\left(\dfrac{2}{x} + y^2\right)}{xz^3\left(\dfrac{1}{z^3}\right)} = \frac{2z^3 + xy^2z^3}{x}$$

Now Try:

4. Simplify, using only positive exponents in the answer.

$$\frac{x^{-1}}{y - x^{-1}}$$

_____

Objective 4 Practice Exercises

For extra help, see Example 4 on pages 388–389 of your text.

Simplify each expression, using only positive exponents in the answer.

10. $\dfrac{4x^{-2}}{2 + 6y^{-3}}$ **10.** _____

11. $\dfrac{s^{-1}+r}{r^{-1}+s}$

11. _____

12. $\dfrac{(m+n)^{-2}}{m^{-2}-n^{-2}}$

12. _____

Chapter 6 RATIONAL EXPRESSIONS AND FUNCTIONS

6.4 Equations with Rational Expressions and Graphs

Learning Objectives
1   Determine the domain of the variable in a rational equation.
2   Solve rational equations.
3   Recognize the graph of a rational function.

Key Terms

Use the vocabulary terms listed below to complete each statement in exercises 1–4.

domain of the variable in a rational equation   **discontinuous**

vertical asymptote   **horizontal asymptote**

1. A rational function in simplest form $f(x) = \dfrac{P(x)}{x-a}$ has the line $x = a$ as a

    _____.

2. The _____ is the intersection of the domains of the rational expressions in the equation.

3. A graph of a function is _____ if there are one or more breaks in the graph.

4. A horizontal line that a graph approaches as |x| gets larger without bound is called a _____.

Objective 1   Determine the domain of the variable in a rational equation.

Video Examples

Review these examples for Objective 1:

1. Find the domain of the variable in each equation.

    a. $\dfrac{4}{5x} + \dfrac{3}{2x} = \dfrac{23}{50}$

    The domains of the three expressions are $\{x \mid x \neq 0\}$, $\{x \mid x \neq 0\}$, and $\{x \mid x \text{ is a real number}\}$. The intersection of these three domains is all real numbers except 0, written in set-builder notation as $\{x \mid x \neq 0\}$.

Now Try:

1. Find the domain of the variable in each equation.

    a. $\dfrac{1}{x} + \dfrac{9}{4} = \dfrac{11}{2x}$

    _____

b. $\dfrac{r+5}{r^2-16} = \dfrac{3}{r-4} + \dfrac{1}{r+4}$.

The domains of the three expressions are respectively, $\{x \mid x \neq \pm 4\}$, $\{x \mid x \neq 4\}$, and $\{x \mid x \neq -4\}$. The domain of the variable is the intersection of the three domains, all real numbers except –4 and 4, written $\{x \mid x \neq \pm 4\}$.

b. $\dfrac{10}{x-7} + \dfrac{7}{x+8} = \dfrac{-3}{x^2+x-56}$

_____

Objective 1 Practice Exercises

For extra help, see Example 1 on page 391 of your text.

(a) Without actually solving the equations below, list all possible numbers that would have to be rejected if they appeared as potential solutions. (b) Then give the domain using set notation.

1. $\dfrac{1}{x} + \dfrac{2}{x+1} = 0$

1. a._____

    b._____

2. $\dfrac{x}{6} - \dfrac{1}{2} = \dfrac{3}{x+1}$

2. a._____

    b._____

3. $\dfrac{3}{3+x} + \dfrac{5}{x^2-x} = \dfrac{9}{x}$

3. a._____

    b._____

Objective 2   Solve rational equations.

Video Examples

Review these examples for Objective 2:

2. Solve $\dfrac{4}{5x} + \dfrac{3}{2x} = \dfrac{23}{50}$.

Step 1 The domain, which excludes 0, was found in Example 1(a).

Step 2 Multiply by the LCD, 50x.

$$50x\left(\frac{4}{5x} + \frac{3}{2x}\right) = 50x\left(\frac{23}{50}\right)$$

Step 3 $50x\left(\dfrac{4}{5x}\right) + 50x\left(\dfrac{3}{2x}\right) = 50x\left(\dfrac{23}{50}\right)$

$$40 + 75 = 23x$$
$$115 = 23x$$
$$5 = x$$

Step 4 Check $\dfrac{4}{5x} + \dfrac{3}{2x} = \dfrac{23}{50}$

$$\frac{4}{5(5)} + \frac{3}{2(5)} \overset{?}{=} \frac{23}{50}$$

$$\frac{4}{25} + \frac{3}{10} \overset{?}{=} \frac{23}{50}$$

$$\frac{23}{50} = \frac{23}{50}$$

The solution set is {5}.

3. Solve $\dfrac{10}{x^2 - 2x} + \dfrac{4}{x} = \dfrac{5}{x - 2}$.

Step 1 The domain excludes 0 and 2.

Step 2 Multiply by the LCD, $x(x - 2)$.

$$x(x-2)\left[\frac{10}{x^2 - 2x} + \frac{4}{x}\right] = x(x-2)\left(\frac{5}{x-2}\right)$$

Step 3

$$x(x-2)\left(\frac{10}{x^2 - 2x}\right) + x(x-2)\left(\frac{4}{x}\right)$$

$$= x(x-2)\left(\frac{5}{x-2}\right)$$

$$10 + 4(x - 2) = 5x$$
$$10 + 4x - 8 = 5x$$
$$2 + 4x = 5x$$
$$2 = x$$

Now Try:

2. Solve $\dfrac{1}{x} + \dfrac{9}{4} = \dfrac{11}{2x}$.

_____

3. Solve $\dfrac{4}{x - 2} - \dfrac{3}{x + 2} = \dfrac{16}{x^2 - 4}$.

_____

Step 4 Since the proposed solution, 2, is not in the domain, it cannot be a solution of the equation.

$$\frac{10}{x^2-2x}+\frac{4}{x}=\frac{5}{x-2}$$

$$\frac{10}{2^2-2(2)}+\frac{4}{2}\overset{?}{=}\frac{5}{2-2}$$

$$\frac{10}{0}+2\overset{?}{=}\frac{5}{0}$$

The equation has no solution The solution set is $\varnothing$.

Objective 2 Practice Exercises

For extra help, see Examples 2–5 on pages 392–394 of your text.

Solve each equation.

4. $\dfrac{8}{2m+4}+\dfrac{2}{3m+6}=\dfrac{7}{9}$

4. _____

5. $\dfrac{1}{b+2}-\dfrac{5}{b^2+9b+14}=\dfrac{-3}{b+7}$

5. _____

6. $\dfrac{-16}{n^2-8n+12}=\dfrac{3}{n-2}+\dfrac{n}{n-6}$

6. _____

Objective 3   Recognize the graph of a rational function.

Video Examples

Review this example for Objective 3:

6. Graph, and give the equations of the vertical and horizontal asymptotes.

$$f(x) = \frac{-2}{x+3}$$

Some ordered pairs that belong to the function are listed below.

x	$f(x) = \dfrac{-2}{x+3}$	x	$f(x) = \dfrac{-2}{x+3}$
-6	$\dfrac{2}{3}$	1	$-\dfrac{1}{2}$
-5	1	2	$-\dfrac{2}{5}$
-4	2	3	$-\dfrac{1}{3}$
-2	-2	4	$-\dfrac{2}{7}$
-1	-1	5	$-\dfrac{1}{4}$
0	$-\dfrac{2}{3}$	6	$-\dfrac{2}{9}$

There is no point on the graph for $x = -3$ because -3 is excluded from the domain of the rational function.

The dashed line $x = -3$ represents the vertical asymptote and is not part of the graph. The graph gets closer to the vertical asymptote as the x-values get closer to -3. Since the x-values approach 0 as $|x|$ increases, the horizontal asymptote is $y = 0$.

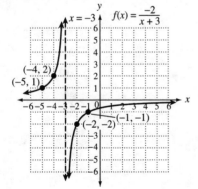

Now Try:

6. Graph, and give the equations of the vertical and horizontal asymptotes.

$$f(x) = -\frac{1}{x}$$

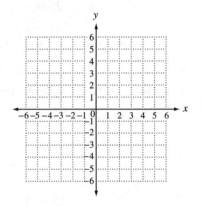

Vertical asymptote: _____

Horizontal asymptote: _____

Name: Date:

Instructor: Section:

Objective 3 Practice Exercises

For extra help, see Example 6 on page 395 of your text.

Graph each rational function. Give the equations of the vertical and horizontal asymptotes.

7. $f(x) = \dfrac{5}{x}$

7.

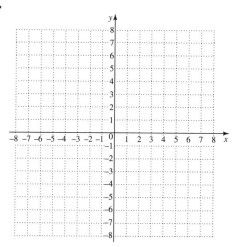

Vertical asymptote: _____

Horizontal asymptote: _____

8. $f(x) = \dfrac{1}{x-4}$

8.

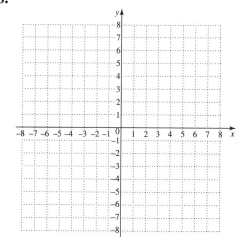

Vertical asymptote: _____

Horizontal asymptote: _____

215

9. $f(x) = \dfrac{3}{x+2}$

9.

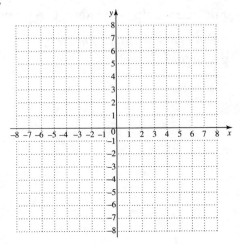

Vertical asymptote: _____

Horizontal asymptote: _____

Chapter 6   RATIONAL EXPRESSIONS AND FUNCTIONS

6.5   Applications of Rational Expressions

Learning Objectives

1	Find the value of an unknown variable in a formula.
2	Solve a formula for a specified variable.
3	Solve applications using proportions.
4	Solve applications about distance, rate, and time.
5	Solve applications about work rates.

Key Terms

Use the vocabulary terms listed below to complete each statement in exercises 1–2.

> ratio   proportion

1. A comparison of two quantities using a quotient is a _____.

2. A statement that two ratios are equal is a _____.

Objective 1   Find the value of an unknown variable in a formula.

Video Examples

Review this example for Objective 1:

1. The height of a trapezoid is given by the formula $h = \dfrac{2A}{B+b}$. Find B, if $A = 40$, $h = 8$ and $b = 3$.

Substitute the given values into the equation and solve for B.

$$h = \frac{2A}{B+b}$$

$$8 = \frac{2(40)}{B+3}$$

$$8(B+3) = \frac{80}{B+3}(B+3)$$

$$8B + 24 = 80$$

$$8B = 56$$

$$B = 7$$

Now Try:

1. The formula for the slope of a line is $m = \dfrac{y_1 - y_2}{x_1 - x_2}$. Find x_2 if $m = 3$, $y_1 = 12$, $y_2 = 8$, and $x_1 = 5$.

_____

Objective 1 Practice Exercises

For extra help, see Example 1 on page 400 of your text.

Find the value of the variable indicated.

1.   If $F = \dfrac{GmM}{d^2}$, $F = 150$, $G = 32$, $M = 50$,   1. _____

     and $d = 10$, find m.

2.   If $\dfrac{1}{f} = \dfrac{1}{d_0} + \dfrac{1}{d_i}$, $f = 10$, and $d_0 = 25$, find d_i.   2. _____

3.   If $c = \dfrac{100b}{L}$, $c = 80$, and $b = 16$, find L.   3. _____

Objective 2   Solve a formula for a specified variable.

Video Examples

Review this example for Objective 2:

2.   Solve the following formula for R_1.

$$\frac{1}{R} = \frac{1}{R_1} + \frac{1}{R_2}$$

Start by multiplying by the LCD, RR_1R_2.

$$RR_1R_2\left(\frac{1}{R}\right) = RR_1R_2\left(\frac{1}{R_1} + \frac{1}{R_2}\right)$$

$$RR_1R_2\left(\frac{1}{R}\right) = RR_1R_2\left(\frac{1}{R_1}\right) + RR_1R_2\left(\frac{1}{R_2}\right)$$

$$R_1R_2 = RR_2 + RR_1$$

$$R_1R_2 - RR_1 = RR_2$$

$$R_1(R_2 - R) = RR_2$$

$$R_1 = \frac{RR_2}{R_2 - R}$$

Now Try:

2.   Solve $\dfrac{1}{R} = \dfrac{1}{R_1} + \dfrac{1}{R_2}$ for R.

_____

Objective 2 Practice Exercises

For extra help, see Examples 2–3 on pages 400–401 of your text.

Solve each formula for the specified variable.

4. $\dfrac{V_1 P_1}{T_1} = \dfrac{V_2 P_2}{T_2}$ for T_2 4. _____

5. $h = \dfrac{2A}{B+b}$ for B 5. _____

6. $E = \dfrac{e(R+r)}{r}$ for r 6. _____

Objective 3   Solve applications using proportions.

Video Examples

Review this example for Objective 3:

5. Ryan's car uses 5 gallons of gasoline to travel 120 miles. If he has 4 gallons of gas in the car, how much more gasoline will he need to travel 288 miles?

    Step 1 Read the problem.

    Step 2 Assign a variable. Let x = the additional number of gallons of gas needed.

    Step 3 Write an equation. Use a proportion.

    $\dfrac{\text{gallons} \rightarrow}{\text{miles} \rightarrow} \dfrac{5}{120} = \dfrac{4+x}{288} \dfrac{\leftarrow \text{gallons}}{\leftarrow \text{miles}}$

    Step 4 Solve.

    $$\frac{5}{120} = \frac{4+x}{288}$$
    $$(288)(5) = (120)(4+x)$$
    $$1440 = 480 + 120x$$
    $$960 = 120x$$
    $$8 = x$$

Now Try:

5. Beth's car uses 12 gallons of gasoline to travel 336 miles. If she has 5 gallons of gas in the car, how much more gasoline will she need to travel 252 miles?

    _____

Step 5 State the answer. Ryan will need 8 more gallons of gas.

Step 6 Check. The original 4 gallons plus 8 gallons equals 12 gallons.

$$\frac{5}{120} \overset{?}{=} \frac{4+8}{288}$$

$$\frac{1}{24} = \frac{1}{24}$$

Objective 3 Practice Exercises

For extra help, see Examples 4–5 on pages 401–402 of your text.

Use proportions to solve each problem.

7.   Ryan's car travels 24 miles using 1 gallon of gas. If Ryan has 4 gallons of gas in his car, how much more gas will he need to travel 288 miles?

7. _____

8.   Connie paid $0.90 in sales tax on a purchase of $12.00. Later that day she purchased an item for which she paid $55.90 including sales tax, at the same tax rate. How much was the sales tax on that item?

8. _____

9.   A certain city with a population of 400,000 had 1600 burglaries during the last year. If the number of burglaries increased by 200 this year and its burglary rate remained the same, by what number did its population increase?

9. _____

Objective 4   Solve applications about distance, rate, and time.

Video Examples

Review this example for Objective 4:

6. Mark can row 5 miles per hour in still water. It takes him as long to row 4 miles upstream as 16 miles downstream. How fast is the current?

Step 1 Read the problem carefully. We must find the speed of the current.

Step 2 Assign a variable. Let x = the speed of the current. Traveling upstream (against the current) slows Mark down, so his rate is the difference between his rate in still water and the rate of the current, that is $5 - x$ mph. Traveling downstream (with the current) speeds him up, so his rate is the sum of his rate in still water and the rate of the current, $5 + x$ mph. We can summarize the given information in a table. Use the formula $d = rt$ or $t = \frac{d}{r}$.

	r	d	t
upstream	$5 - x$	4	$\dfrac{4}{5-x}$
downstream	$5 + x$	16	$\dfrac{16}{5+x}$

Step 3 Write an equation. The times are equal so we have $\dfrac{4}{5-x} = \dfrac{16}{5+x}$.

Step 4 Solve.
$$\frac{4}{5-x} = \frac{16}{5+x}$$
$$(5-x)(5+x)\left(\frac{4}{5-x}\right) = (5-x)(5+x)\left(\frac{16}{5+x}\right)$$
$$4(5+x) = (5-x)16$$
$$20 + 4x = 80 - 16x$$
$$20x = 60$$
$$x = 3$$

Step 5 State the answer.
The speed of the current is 3 miles per hour.

Step 6 Check the solution in the words of the original problem. Mike can row upstream $5 - 3 = 2$ miles per hour. It will take him $\frac{4}{2} = 2$

Now Try:

6. A boat goes 6 miles per hour in still water. It takes as long to go 40 miles upstream as 80 miles downstream. Find the speed of the current.

_____

hours to row four miles upstream. Mike can row downstream $5 + 3 = 8$ miles per hour. It will take him $\frac{16}{8} = 2$ hours to row 16 miles downstream.

The time upstream equals the time downstream, as required.

Objective 4 Practice Exercises

For extra help, see Examples 6–7 on pages 403–405 of your text.

Solve each problem.

10. Lauren's boat can go 9 miles per hour in still water. How far downstream can Lauren go if the river has a current of 3 miles per hour and she must be back in 4 hours.

10.

11. Pauline and Pete agree to meet in Columbia. Pauline travels 120 miles, while Pete travels 80 miles. If Pauline's speed is 20 miles per hour greater than Pete's and they both spend the same amount of time traveling, at what speed does each travel?

11.

12. Olivia can ride her bike 4 miles per hour faster than Ted can ride his bike. If Olivia can go 30 miles in the same time that Ted can go 15 miles, what are their speeds?

12.

Name: _____ Date: _____

Instructor: _____ Section: _____

Objective 5   Solve applications about work rates.

Video Examples

Review this example for Objective 5:

8.   One pipe can fill a swimming pool in 8 hours and another pipe can fill the pool in 12 hours. How long will it take to fill the pool if both pipes are open?

Let x = the number of hours working together. The rate for the first pipe is $\frac{1}{8}$.

The rate for the second pipe is $\frac{1}{12}$.

The sum of the fractional part for each pipe multiplied by the time working together is the whole job.

$$\frac{1}{8}x + \frac{1}{12}x = 1$$

$$48\left(\frac{1}{8}x + \frac{1}{12}x\right) = 48(1)$$

$$48\left(\frac{1}{8}x\right) + 48\left(\frac{1}{12}x\right) = 48$$

$$6x + 4x = 48$$

$$10x = 48$$

$$x = \frac{48}{10} = \frac{24}{5} = 4\frac{4}{5}$$

Working together, it takes $4\frac{4}{5}$ hours to fill the pool. A check confirms this answer.

Now Try:

8.   Chuck can weed the garden in $\frac{1}{2}$ hour, but David takes 2 hours. How long does it take them to weed the garden if they work together?

_____

Objective 5 Practice Exercises

For extra help, see Example 8 on pages 405–406 of your text.

Solve each problem.

13.   Kelly can clean the house in 6 hours, but it takes Linda 4 hours. How long would it take them to clean the house if they worked together?

13.   _____

14. A swimming pool can be filled by an inlet pipe in 18   **14.** _____
hours and emptied by an outlet pipe in 24 hours.
How long will it take to fill the empty pool if the
outlet pipe is accidentally left open at the same time
as the inlet pipe is opened?

15. Fred can seal an asphalt driveway in $\frac{1}{3}$ the time it   **15.** _____
takes John. Working together, it takes them $1\frac{1}{2}$
hours. How long would it have taken Fred working
alone?

Chapter 6   RATIONAL EXPRESSIONS AND FUNCTIONS

6.6   Variation

Learning Objectives
1   Write an equation expressing direct variation.
2   Find the constant of variation, and solve direct variation problems.
3   Solve inverse variation problems.
4   Solve joint variation problems.
5   Solve combined variation problems.

Key Terms

Use the vocabulary terms listed below to complete each statement in exercises 1−3.

varies directly   **varies inversely**   **constant of variation**

1.   In the equations for direct and inverse variation, k is the _____.

2.   If there exists a real number k such that $y = \dfrac{k}{x}$, then y _____

      as x.

3.   If there exists a real number k such that $y = kx$, then y _____

      as x.

Objective 1   Write an equation expressing direct variation.

For extra help, see pages 411–412 of your text.

Objective 2   Find the constant of variation, and solve direct variation problems.

Video Examples

Review these examples for Objective 2:

1. If 12 gallons of gasoline cost $34.68, how much does 1 gallon of gasoline cost? Write the variation equation.

   Let g represent the number of gallons of gasoline and let C represent the total cost of the gasoline. Then the variation equation is $C = kg$.

   $$C = kg$$

   $$34.68 = 12k$$

   $$2.89 = k$$

   The cost per gallon is $2.89.
   The variation equation is $C = 2.89g$.

Now Try:

1. One week a manufacturer sold 1200 items for a total profit of $30,000. What was the profit for one item. Write the variation equation.

   _____

   _____

2. A person's weight on the moon varies directly with the person's weight on Earth. A 120-pound person would weigh about 20 pounds on the moon. How much would a 150-pound person weigh on the moon?

If m represents the person's weight on the moon and w represents the person's weight on Earth.

$$m = kw$$

$20 = k \cdot 120$ Let $m = 20$ and $w = 120$.

$\dfrac{20}{120} = \dfrac{1}{6} = k$ Solve for k; lowest terms

Now, substitute $\dfrac{1}{6}$ for k and 150 for w in the variation equation.

$$m = kw$$

$$m = \frac{1}{6} \cdot 150 = 25$$

A 150-pound person will weigh 25 pounds on the moon.

3. The surface area of a sphere varies directly as the square of its radius. If the surface area of a sphere with a radius of 12 inches is 576π square inches, find the surface area of a sphere with a radius of 3 inches.

Step 1 A represents the surface area and r represents the radius.

$$A = kr^2$$

Step 2 Find the value of k when A is 576π and r is 12. $A = kr^2$

$$576\pi = k \cdot 12^2$$

$$576\pi = 144k$$

$$\frac{576\pi}{144} = 4\pi = k$$

Step 3 Rewrite the variation equation.

$$A = 4\pi r^2$$

Step 4 Let $r = 3$ to find the surface area.

$$A = 4\pi \cdot 3^2 = 4\pi \cdot 9 = 36\pi$$

The surface area of a sphere with a radius of 3 inches is 36π square inches.

2. The pressure exerted by a certain liquid at a given point varies directly as the depth of the point beneath the surface of the liquid. The pressure at 10 feet is 50 pounds per square inch (psi). What is the pressure at 25 feet?

_____

3. The area of a circle varies directly as the square of the radius. A circle with a radius of 5 centimeters has an area of 78.5 square centimeters. Find the area if the radius changes to 7 centimeters.

_____

Objective 2 Practice Exercises

For extra help, see Examples 1–3 on pages 412–414 of your text.

Find the constant of variation, and write a direct variation equation.

1. $y = 13.75$ when $x = 55$

1. _____

Solve each problem.

2.   The circumference of a circle varies directly as the radius. A circle with a radius of 7 centimeters has a circumference of 43.96 centimeters. Find the circumference of the circle if the radius changes to 11 centimeters.

2. _____

3.   The force required to compress a spring varies directly as the change in length of the spring. If a force of 20 newtons is required to compress a spring 2 centimeters in length, how much force is required to compress a spring of length 10 centimeters?

3. _____

Objective 3   Solve inverse variation problems.

Video Examples

Review these examples for Objective 3:

4.   For a specified distance, time varies inversely with speed. If Ramona walks a certain distance on a treadmill in 40 minutes at 4.2 miles per hour, how long will it take her to walk the same distance at 3.5 miles per hour?

   Let t = time and s = speed.
   Since t varies inversely as s, there is a constant k such that $t = \dfrac{k}{s}$. Recall that $40 \text{ min} = \dfrac{40}{60} \text{ hr}$.

$$t = \frac{k}{s}$$
$$\frac{40}{60} = \frac{k}{4.2}$$
$$2.8 = k$$

Now Try:

4.   The length of a violin string varies inversely with the frequency of its vibrations. A 10-inch violin string vibrates at a frequency of 512 cycles per second. Find the frequency of an 8-inch string.

_____

Now use $t = \dfrac{k}{s}$ to find the value of t

when $s = 3.5$.

$$t = \dfrac{2.8}{3.5} = \dfrac{4}{5}$$

It takes $\dfrac{4}{5}$ hr, or 48 min to walk the same

distance.

5. With constant power, the resistance used in a simple electric circuit varies inversely as the square of the current. If the resistance is 120 ohms when the current is 12 amps, find the resistance if the current is reduced to 9 amps.

Let R represent resistance (in ohms) and

I = current (in amps). Then $R = \dfrac{k}{I^2}$.

First, we solve for the constant of variation by substituting 120 for R and 12 for I.

$$120 = \dfrac{k}{12^2}$$

$$k = 120 \cdot 12^2$$

Now use the value for k and 9 for I to find R.

$$R = \dfrac{120 \cdot 12^2}{9^2} \approx 213.3$$

The resistance is about 213.3 ohms when the current is 9 amps.

5. If y varies inversely as x^3, and $y = 9$ when $x = 2$, find y when $x = 4$.

5. _____

Objective 3 Practice Exercises

For extra help, see Examples 4–5 on page 415 of your text.

Solve each problem.

4. The illumination produced by a light source varies inversely as the square of the distance from the source. If the illumination produced 4 feet from a light source is 75 footcandles, find the illumination produced 9 feet from the same source.

4. _____

5.   The weight of an object varies inversely as the square of its distance from the center of Earth. If an object 8000 miles from the center of Earth weighs 90 pounds, find its weight when it is 12,000 miles from the center of Earth.

5. _____

6.   The speed of a pulley varies inversely as its diameter. One kind of pulley, with a diameter of 3 inches, turns at 150 revolutions per minute. Find the speed of a similar pulley with diameter of 5 inches.

6. _____

Objective 4   Solve joint variation problems.

Video Examples

Review this example for Objective 4:

6. For a fixed interest rate, interest varies jointly as the principal and the time in years. If $5000 invested for 4 years earns $900, how much interest will $6000 invested for 3 years earn at the same interest rate?

Let I = the interest, p = the principal, and t = the time in years. Then, $I = kpt$.

$$I = kpt$$
$$900 = k \cdot 5000 \cdot 4 \quad \text{Substitute given values.}$$
$$\frac{900}{20,000} = \frac{9}{200} = k$$

Now use $k = \frac{9}{200}$.

$$I = \frac{9}{200} \cdot 6000 \cdot 3 = 810$$

$6000 invested for three years will earn $810 in interest.

Now Try:

6. The strength of a rectangular beam varies jointly as its width and the square of its depth. If the strength of a beam 2 inches wide by 10 inches deep is 1000 pounds per square inch, what is the strength of a beam 4 inches wide and 8 inches deep?

_____

Objective 4 Practice Exercises

For extra help, see Example 6 on page 416 of your text.

Solve each problem.

7. Suppose d varies jointly as f^2 and g^2, and $d = 384$ when $f = 3$ and $g = 8$. Find d when $f = 6$ and $g = 2$.

7. _____

8. The work w (in joules) done when lifting an object is jointly proportional to the product of the mass m (in kg) of the object and the height h (in meter) the object is lifted. If the work done when a 120 kg object is lifted 1.8 meters above the ground is 2116.8 joules, how much work is done when lifting a 100kg object 1.5 meters above the ground?

8. _____

9. The absolute temperature of an ideal gas varies jointly as its pressure and its volume. If the absolute temperature is 250° when the pressure is 25 pounds per square centimeter and the volume is 50 cubic centimeters, find the absolute temperature when the pressure is 50 pounds per square centimeter and the volume is 75 cubic centimeters.

9. _____

Objective 5   Solve combined variation problems.

Video Examples

Review this example for Objective 5:

7. The number of hours h that it takes w workers to assemble x machines varies directly as the number of machines and inversely as the number of workers. If four workers can assemble 12 machines in four hours, how many workers are needed to assemble 36 machines in eight hours?

The variation equation is $h = \dfrac{kx}{w}$.

To find k, let $h = 4$, $x = 12$, and $w = 4$.

$$4 = \frac{k \cdot 12}{4}$$

$$k = \frac{4 \cdot 4}{12}$$

$$k = \frac{4}{3}$$

Now find w when $x = 36$ and $h = 8$.

$$8 = \frac{\frac{4}{3} \cdot 36}{w}$$

$$8w = 48$$

$$w = 6$$

Eight workers are needed to assemble 36 machines in eight hours.

Now Try:

7. The volume of a gas varies directly as its temperature and inversely as its pressure. The volume of a gas at 85° C at a pressure of 12 kg/cm^2 is 300 cm^3. What is the volume when the pressure is 20 kg/cm^2 and the temperature is 30° C?

_____

Objective 5 Practice Exercises

For extra help, see Example 7 on pages 416–417 of your text.

Solve each problem.

10. The volume of a gas varies inversely as the pressure and directly as the temperature. If a certain gas occupies a volume of 1.3 liters at 300 K and a pressure of 18 kilograms per square centimeter, find the volume at 340 K and a pressure of 24 kilograms per square centimeter.

10. _____

11. The time required to lay a sidewalk varies directly as its length and inversely as the number of people who are working on the job. If three people can lay a sidewalk 100 feet long in 15 hours, how long would it take two people to lay a sidewalk 40 feet long?

11. _____

12. When an object is moving in a circular path, the centripetal force varies directly as the square of the velocity and inversely as the radius of the circle. A stone that is whirled at the end of a string 50 centimeters long at 900 centimeters per second has a centripetal force of 3,240,000 dynes. Find the centripetal force if the stone is whirled at the end of a string 75 centimeters long at 1500 centimeters per second.

12. _____

Chapter 7   ROOTS, RADICALS, AND ROOT FUNCTIONS

7.1   Radical Expressions and Graphs

Learning Objectives
1   Find roots of numbers.
2   Find principal roots.
3   Graph functions defined by radical expressions.
4   Find nth roots of nth powers.
5   Use a calculator to find roots.

Key Terms

Use the vocabulary terms listed below to complete each statement in exercises 1−8.

square root	principal square root	radicand
radical	radical expression	perfect square
cube root	index (order)	

1.   The number or expression inside a radical sign is called the _____.

2.   A number with a rational square root is called a _____.

3.   In a radical of the form $\sqrt[n]{a}$, the number n is the _____.

4.   The number b is a _____ of a if $b^2 = a$.

5.   The expression $\sqrt[n]{a}$ is called a _____.

6.   The positive square root of a number is its _____.

7.   A _____ is a radical sign and the number or expression in it.

8.   The number b is a _____ of a if $b^3 = a$.

Objective 1   Find roots of numbers.

Video Examples

Review these examples for Objective 1:

1. Simplify.

    a. $\sqrt[4]{81}$

    $\sqrt[4]{81} = 3$, because $3^4 = 81$.

Now Try:

1. Simplify.

    a. $\sqrt[4]{256}$

    _____

b. $\sqrt[3]{\dfrac{125}{343}}$

b. $\sqrt[3]{\dfrac{64}{27}}$

_____

$$\sqrt[3]{\dfrac{125}{343}} = \dfrac{5}{7}, \text{ because } \left(\dfrac{5}{7}\right)^3 = \dfrac{125}{343}.$$

Objective 1 Practice Exercises

For extra help, see Example 1 on page 434 of your text.

Find each root that is a real number. Use a calculator as necessary.

1. $\sqrt[4]{625}$

1. _____

2. $\sqrt[6]{64}$

2. _____

3. $\sqrt[4]{6561}$

3. _____

Objective 2   Find principal roots.

Video Examples

Review these examples for Objective 2:

2. Find each root.

   a. $-\sqrt{144}$

   $-\sqrt{144} = -12$

   b. $\sqrt[3]{-64}$

   $\sqrt[3]{-64} = -4$, because $(-4)^3 = -64$.

   c. $\sqrt[4]{-256}$

   The index is even and the radicand is negative, so $\sqrt[4]{-256}$ is not a real number.

Now Try:

2. Find each root.

   a. $-\sqrt{121}$

   _____

   b. $\sqrt[3]{-125}$

   _____

   c. $\sqrt[4]{-625}$

   _____

Objective 2 Practice Exercises

For extra help, see Example 2 on page 435 of your text.

Find each root.

4. $\sqrt[4]{-81}$

4. _____

5. $\sqrt[4]{256}$

5. _____

6. $\sqrt[7]{-1}$

6. _____

Name: Date:

Instructor: Section:

Objective 3   Graph functions defined by radical expressions.

Video Examples

Review these examples for Objective 3:

3. Graph each function by creating a table of values. Give the domain and the range.

    a. $f(x) = \sqrt{x-1}$

Create a table of values.

x	$f(x) = \sqrt{x-1}$
1	$\sqrt{1-1} = 0$
5	$\sqrt{5-1} = 2$
10	$\sqrt{10-1} = 3$

For the radicand to be nonnegative, we must have $x - 1 \geq 0$ or $x \geq 1$. Therefore, the domain is $[1, \infty)$. Function values are nonnegative, so the range is $[0, \infty)$.

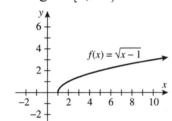

    b. $f(x) = \sqrt[3]{x} + 1$

Create a table of values.

x	$f(x) = \sqrt[3]{x} + 1$
-8	$\sqrt[3]{-8} + 1 = -1$
-1	$\sqrt[3]{-1} + 1 = 0$
0	$\sqrt[3]{0} + 1 = 1$
1	$\sqrt[3]{1} + 1 = 2$
8	$\sqrt[3]{8} + 1 = 3$

Both the domain and range are $(-\infty, \infty)$.

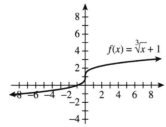

Now Try:

3. Graph each function by creating a table of values. Give the domain and the range.

    a. $f(x) = \sqrt{x-1}$

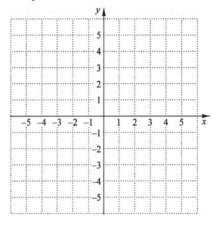

    domain: _____

    range: _____

    b. $f(x) = \sqrt[3]{x} + 1$

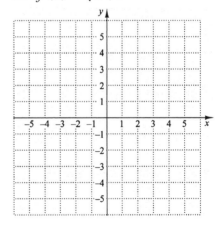

    domain: _____

    range: _____

Name: Date:

Instructor: Section:

Objective 3 Practice Exercises

For extra help, see Example 3 on page 436 of your text.

Graph each function and give its domain and its range.

7. $f(x) = \sqrt{x} + 2$ **7.**

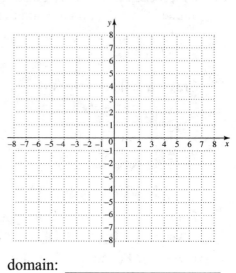

domain: _____

range: _____

8. $f(x) = \sqrt[3]{x} - 2$ **8.**

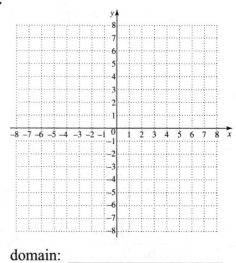

domain: _____

range: _____

9. $f(x) = \sqrt[3]{x} + 2$

9.

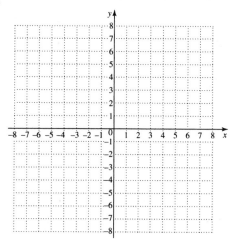

domain: _____

range: _____

Objective 4   Find *n*th roots of *n*th powers.

Video Examples

Review these examples for Objective 4:

4. Find each square root. In part (d), *m* is a real number.

a. $\sqrt{33^2}$

$\sqrt{33^2} = |33| = 33$

b. $\sqrt{(-m)^2}$

$\sqrt{(-m)^2} = |-m| = |m|$

5. Simplify each root.

a. $\sqrt[4]{(-5)^4}$

n is even. Use absolute value.

$\sqrt[4]{(-5)^4} = |-5| = 5$

b. $\sqrt[5]{s^{20}}$

$\sqrt[5]{s^{20}} = s^4$, because $s^{20} = (s^4)^5$.

Now Try:

4. Find each square root. In part (d), *n* is a real number.

a. $\sqrt{73^2}$

_____

b. $\sqrt{(-n)^2}$

_____

5. Simplify each root.

a. $\sqrt[8]{(-4)^8}$

_____

b. $\sqrt[3]{w^{30}}$

_____

Objective 4 Practice Exercises

For extra help, see Examples 4–5 on page 437 of your text.

Simplify each root.

10. $\sqrt{(-9)^2}$

10. _____

11. $-\sqrt[5]{x^5}$

11. _____

12. $-\sqrt[4]{x^{16}}$

12. _____

Objective 5   Use a calculator to find roots.

Video Examples

Review this example for Objective 5:

7.   The minimum speed at which a driver is traveling when a car skids to a stop can be estimated by the formula $s = \sqrt{30fd}$, where s is the speed of the car in miles per hour when the brakes are applied, f is a constant drag factor, and d is the length of the skid marks in feet. Suppose that a car skids to a stop leaving a skid of 90 feet. Assume that the drag factor is 0.7. How fast was the driver going? Give your answer to the nearest tenth.

Start with the given formula. Substitute for f and d, and use a calculator.

$$s = \sqrt{30fd}$$
$$s = \sqrt{30(0.7)(90)}$$
$$s \approx 43.5$$

The driver was traveling at about 43.5 miles per hour at the start of the skid.

Now Try:

7.   Use the formula at the left to find out how fast a driver was traveling if the skid mark is 20 feet and the drag factor is 0.9. Give your answer to the nearest tenth.

_____

Objective 5 Practice Exercises

For extra help, see Examples 6–7 on page 438 of your text.

Use a calculator to find a decimal approximation for each radical. Give the answer to the nearest thousandth.

13. $\sqrt[3]{701}$ 13. _____

14. $-\sqrt{990}$ 14. _____

15.   The time t in seconds for one complete swing of a   15. _____
      simple pendulum, where L is the length of the
      pendulum in feet is $t = 2\pi\sqrt{\dfrac{L}{32}}$. Find the time of a
      complete swing of a 4-ft pendulum to the nearest
      tenth of a second.

Chapter 7 ROOTS, RADICALS, AND ROOT FUNCTIONS

7.2   Rational Exponents

Learning Objectives
1   Use exponential notation for nth roots.
2   Define and use expressions of the form $a^{m/n}$.
3   Convert between radicals and rational exponents.
4   Use the rules for exponents with rational exponents.

Key Terms

Use the vocabulary terms listed below to complete each statement in exercises 1–3.

product rule for exponents   **quotient rule for exponents**

power rule for exponents

1. $\left(x^2 y^3\right)^4 = x^8 y^{12}$ is an example of the _____.

2. $w^5 w^3 = w^8$ is an example of the _____.

3. $\dfrac{z^6}{z^4} = z^2$ is an example of the _____.

Objective 1   Use exponential notation for nth roots.

Video Examples

Review these examples for Objective 1:
1.   Evaluate each exponential.

a. $25^{1/2}$

$25^{1/2} = \sqrt{25} = 5$

b. $(-8)^{1/3}$

$(-8)^{1/3} = \sqrt[3]{-8} = -2$

c. $(-81)^{1/4}$

$(-81)^{1/4} = \sqrt[4]{-81}$ is not a real number.

Now Try:
1.   Evaluate each exponential.

a. $121^{1/2}$

_____

b. $(-243)^{1/5}$

_____

c. $(-1024)^{1/10}$

_____

Objective 1 Practice Exercises

For extra help, see Example 1 on page 442 of your text.

Evaluate each exponential.

1. $-256^{1/4}$

1. _____

2. $16^{1/2}$

2. _____

3. $(-3375)^{1/3}$

3. _____

Objective 2   Define and use expressions of the form $a^{m/n}$.

Video Examples

Review these examples for Objective 2:

2. Evaluate each polynomial.

   a. $100^{3/2}$

   $100^{3/2} = (100^{1/2})^3 = 10^3 = 1000$

   b. $-729^{5/6}$

   $-729^{5/6} = -(729^{1/6})^5 = -(3)^5 = -243$

3. Evaluate the exponential.

   $625^{-3/4}$

   $625^{-3/4} = \dfrac{1}{625^{3/4}} = \dfrac{1}{(625^{1/4})^3} = \dfrac{1}{\left(\sqrt[4]{625}\right)^3}$

   $= \dfrac{1}{5^3} = \dfrac{1}{125}$

Now Try:

2. Evaluate each polynomial.

   a. $27^{2/3}$

   _____

   b. $-36^{3/2}$

   _____

3. Evaluate the exponential.

   $32^{-2/5}$

   _____

Objective 2 Practice Exercises

For extra help, see Examples 2–3 on pages 443–444 of your text.

Evaluate each exponential.

4. $-81^{5/4}$

4. _____

5. $36^{5/2}$

5. _____

6. $\left(\dfrac{125}{27}\right)^{-2/3}$

6. _____

Objective 3   Convert between radicals and rational exponents.

Video Examples

Review these examples for Objective 3:

4. Write each radical as an exponential and simplify. Assume that all variables represent positive real numbers. Use the definition that takes the root first.

a. $17^{1/2}$

$17^{1/2} = \sqrt{17}$

b. $2x^{2/5} - (4x)^{5/6}$

$2x^{2/5} - (4x)^{5/6} = 2\left(\sqrt[5]{x}\right)^2 - \left(\sqrt[6]{4x}\right)^5$

c. $\sqrt[3]{3^6}$

$\sqrt[3]{3^6} = 3^{6/3} = 3^2 = 9$

Now Try:

4. Write each radical as an exponential and simplify. Assume that all variables represent positive real numbers. Use the definition that takes the root first.

a. $23^{1/3}$

_____

b. $(2x)^{4/3} - 3x^{2/5}$

_____

c. $\sqrt{10^4}$

_____

Objective 3 Practice Exercises

For extra help, see Example 4 on page 445 of your text.

Write with radicals. Assume that all variables represent positive real numbers.

7. $4y^{2/5} + (5x)^{1/5}$

7. _____

8. $\left(2x^4 - 3y^2\right)^{-4/3}$

8. _____

Simplify the radical by rewriting it with a rational exponent. Write answer in radical form if necessary. Assume that variables represent positive real numbers.

9. $\sqrt[8]{a^2}$

9. _____

Objective 4   Use the rules for exponents with rational exponents.

Video Examples

Review these examples for Objective 4:

5. Write with only positive exponents. Assume that all variables represent positive real numbers.

a. $13^{4/5} \cdot 13^{1/2}$

$$13^{4/5} \cdot 13^{1/2} = 13^{4/5+1/2} = 13^{13/10}$$

b. $\left(\dfrac{c^6 x^3}{c^{-2} x^{1/2}} \right)^{-3/4}$

$$\left(\dfrac{c^6 x^{1/2}}{c^{-2} x^3} \right)^{-3/4} = \left(c^{6-(-2)} x^{1/2-3} \right)^{-3/4}$$

$$= \left(c^8 x^{-5/2} \right)^{-3/4}$$

$$= \left(c^8 \right)^{-3/4} \left(x^{-5/2} \right)^{-3/4}$$

$$= c^{-6} x^{15/8}$$

$$= \dfrac{x^{15/8}}{c^6}$$

Now Try:

5. Write with only positive exponents. Assume that all variables represent positive real numbers.

a. $5^{3/4} \cdot 5^{7/4}$

_____

b. $\left(\dfrac{x^{-1} y^{2/3}}{x^{1/3} y^{1/2}} \right)^{-3/2}$

_____

6. Write all radicals as exponentials, and then apply the rules for rational exponents. Leave answers in exponential form. Assume that all variables represent positive real numbers.

   a. $\sqrt[4]{x^3} \cdot \sqrt[5]{x}$

   $$\sqrt[4]{x^3} \cdot \sqrt[5]{x} = x^{3/4} \cdot x^{1/5}$$
   $$= x^{3/4+1/5}$$
   $$= x^{15/20+4/20}$$
   $$= x^{19/20}$$

   b. $\sqrt{\sqrt[4]{y^3}}$

   $$\sqrt{\sqrt[4]{y^3}} = \sqrt{y^{3/4}} = \left(y^{3/4}\right)^{1/2} = y^{3/8}$$

6. Write all radicals as exponentials, and then apply the rules for rational exponents. Leave answers in exponential form. Assume that all variables represent positive real numbers.

   a. $\sqrt[6]{x^3} \cdot \sqrt[3]{x^2}$

   _____

   b. $\sqrt[3]{\sqrt[4]{x^3}}$

   _____

Objective 4 Practice Exercises

For extra help, see Examples 5–6 on pages 446–447 of your text.

Use the rules of exponents to simplify each expression. Write all answers with positive exponents. Assume that variables represent positive real numbers.

10. $y^{7/3} \cdot y^{-4/3}$ 10. _____

11. $\dfrac{a^{2/3} \cdot a^{-1/3}}{\left(a^{-1/6}\right)^3}$ 11. _____

12. $\dfrac{\left(x^{-3}y^2\right)^{2/3}}{\left(x^2y^{-5}\right)^{2/5}}$ 12. _____

Chapter 7 ROOTS, RADICALS, AND ROOT FUNCTIONS

7.3   Simplifying Radicals, the Distance Formula, and Circles

Learning Objectives
1   Use the product rule for radicals.
2   Use the quotient rule for radicals.
3   Simplify radicals.
4   Simplify products and quotients of radicals with different indexes.
5   Use the Pythagorean theorem.
6   Use the distance formula.
7   Find an equation of a circle given its center and radius.

Key Terms

Use the vocabulary terms listed below to complete each statement in exercises 1−6.

      index   radicand   hypotenuse   legs

      radius   circle   center

1.   In a right triangle, the side opposite the right angle is called the

    _____.

2.   In the expression $\sqrt[4]{x^2}$, the "4" is the _____ and x^2 is
    the _____.

3.   In a right triangle, the sides that form the right angle are called the

    _____.

4.   A(n) _____ is the set of all points in a plane that lie a
    fixed distance from a fixed point.

5.   A fixed point such that every point on a circle is a fixed distance from it is the

    _____.

6.   The distance from the center of a circle to a point on the circle is called the

    _____.

Objective 1   Use the product rule for radicals.

Video Examples

Review these examples for Objective 1:

1. Multiply. Assume that all variables represent positive real numbers.

    a. $\sqrt{13} \cdot \sqrt{5}$

    $\sqrt{13} \cdot \sqrt{5} = \sqrt{13 \cdot 5} = \sqrt{65}$

Now Try:

1. Multiply. Assume that all variables represent positive real numbers.

    a. $\sqrt{2} \cdot \sqrt{7}$

    _____

b. $\sqrt{5x} \cdot \sqrt{2yz}$

$\sqrt{5x} \cdot \sqrt{2yz} = \sqrt{10xyz}$

2. Multiply. Assume that all variables represent positive real numbers.

    a. $\sqrt[4]{2} \cdot \sqrt[4]{2x}$

$\sqrt[4]{2} \cdot \sqrt[4]{2x} = \sqrt[4]{2 \cdot 2x} = \sqrt[4]{4x}$

    b. $\sqrt[3]{8x} \cdot \sqrt[3]{2y^2}$

$\sqrt[3]{8x} \cdot \sqrt[3]{2y^2} = \sqrt[3]{8x \cdot 2y^2} = \sqrt[3]{16xy^2}$

    c. $\sqrt[5]{6r^2} \cdot \sqrt[5]{4r^2}$

$\sqrt[5]{6r^2} \cdot \sqrt[5]{4r^2} = \sqrt[5]{6r^2 \cdot 4r^2} = \sqrt[5]{24r^4}$

    d. $\sqrt[5]{2} \cdot \sqrt[4]{6}$

$\sqrt[5]{2} \cdot \sqrt[4]{6}$ cannot be simplified using the product rule for radicals, because the indexes (5 and 4) are different.

b. $\sqrt{3} \cdot \sqrt{11mn}$

_____

2. Multiply. Assume that all variables represent positive real numbers.

    a. $\sqrt[3]{3} \cdot \sqrt[3]{7}$

_____

    b. $\sqrt[3]{7x} \cdot \sqrt[3]{5y}$

_____

    c. $\sqrt[5]{4w} \cdot \sqrt[5]{2w^3}$

_____

    d. $\sqrt{3} \cdot \sqrt[3]{64}$

_____

Objective 1 Practice Exercises

For extra help, see Examples 1–2 on page 451 of your text.

Multiply. Assume that variables represent positive real numbers.

1. $\sqrt{7x} \cdot \sqrt{6t}$ **1.** _____

2. $\sqrt[5]{6r^2t^3} \cdot \sqrt[5]{4r^2t}$ **2.** _____

3. $\sqrt{3} \cdot \sqrt[3]{7}$ **3.** _____

Name: Date:

Instructor: Section:

Objective 2   Use the quotient rule for radicals.

Video Examples

Review these examples for Objective 2:

3. Simplify. Assume that all variables represent positive real numbers.

     a. $\sqrt{\dfrac{64}{9}}$

$$\sqrt{\dfrac{64}{9}} = \dfrac{\sqrt{64}}{\sqrt{9}} = \dfrac{8}{3}$$

     b. $\sqrt{\dfrac{5}{16}}$

$$\sqrt{\dfrac{5}{16}} = \dfrac{\sqrt{5}}{\sqrt{16}} = \dfrac{\sqrt{5}}{4}$$

     c. $\sqrt[3]{-\dfrac{27}{8}}$

$$\sqrt[3]{-\dfrac{27}{8}} = \dfrac{\sqrt[3]{-27}}{\sqrt[3]{8}} = \dfrac{-3}{2} = -\dfrac{3}{2}$$

     d. $\sqrt[5]{-\dfrac{a^3}{243}}$

$$\sqrt[5]{-\dfrac{a^3}{243}} = \dfrac{\sqrt[5]{a^3}}{\sqrt[5]{-243}} = \dfrac{\sqrt[5]{a^3}}{-3} = -\dfrac{\sqrt[5]{a^3}}{3}$$

     e. $\sqrt{\dfrac{z^4}{36}}$

$$\sqrt{\dfrac{z^4}{36}} = \dfrac{\sqrt{z^4}}{\sqrt{36}} = \dfrac{z^2}{6}$$

Now Try:

3. Simplify. Assume that all variables represent positive real numbers.

     a. $\sqrt{\dfrac{36}{49}}$

     _____

     b. $\sqrt{\dfrac{13}{81}}$

     _____

     c. $\sqrt[3]{-\dfrac{343}{125}}$

     _____

     d. $\sqrt[3]{-\dfrac{a^6}{125}}$

     _____

     e. $\sqrt[4]{\dfrac{m}{81}}$

     _____

Objective 2 Practice Exercises

For extra help, see Example 3 on page 451 of your text.

Simplify each radical. Assume that variables represent positive real numbers.

4. $\sqrt[3]{\dfrac{27}{8}}$

                                                   4. _____

5. $\sqrt[5]{\dfrac{7x}{32}}$ **5.** _____

6. $\sqrt[3]{-\dfrac{x^9}{216}}$ **6.** _____

Objective 3   Simplify radicals.

Video Examples

Review these examples for Objective 3:   **Now Try:**

4.   Simplify.   **4.**   Simplify.

    a. $\sqrt{90}$ **a.** $\sqrt{84}$

$$\sqrt{90} = \sqrt{9 \cdot 10}$$
$$= \sqrt{9} \cdot \sqrt{10}$$
$$= 3\sqrt{10}$$
_____

    b. $\sqrt{288}$ **b.** $\sqrt{162}$

$$\sqrt{288} = \sqrt{144 \cdot 2}$$
$$= \sqrt{144} \cdot \sqrt{2}$$
$$= 12\sqrt{2}$$
_____

    c. $\sqrt{35}$ **c.** $\sqrt{95}$

No perfect square (other than 1) divides into 35, so $\sqrt{35}$ cannot be simplified further.
_____

    d. $\sqrt[3]{81}$ **d.** $\sqrt[3]{256}$

$$\sqrt[3]{81} = \sqrt[3]{27 \cdot 3} = \sqrt[3]{27} \cdot \sqrt[3]{3} = 3\sqrt[3]{3}$$
_____

    e. $-\sqrt[4]{3125}$ **e.** $-\sqrt[5]{512}$

$$-\sqrt[4]{3125} = -\sqrt[4]{5^5} = \sqrt[4]{5^4 \cdot 5}$$
$$= -\sqrt[4]{5^4} \cdot \sqrt[4]{5}$$
$$= -5\sqrt[4]{5}$$
_____

5. Simplify. Assume that all variables represent positive real numbers.

a. $\sqrt{81x^3}$

$$\sqrt{81x^3} = \sqrt{9^2 \cdot x^2 \cdot x} = 9x\sqrt{x}$$

b. $\sqrt{56x^7y^6}$

$$\sqrt{56x^7y^6} = \sqrt{4 \cdot 14 \cdot (x^3)^2 \cdot x \cdot (y^3)^2}$$
$$= 2x^3y^3\sqrt{14x}$$

c. $\sqrt[3]{-270b^4c^8}$

$$\sqrt[3]{-270b^4c^8} = \sqrt[3]{(-27b^3c^6)(10bc^2)}$$
$$= \sqrt[3]{-27b^3c^6} \cdot \sqrt[3]{10bc^2}$$
$$= -3bc^2\sqrt[3]{10bc^2}$$

d. $-\sqrt[6]{448a^7b^7}$

$$-\sqrt[6]{448a^7b^7} = -\sqrt[6]{(64a^6b^6)(7ab)}$$
$$= -\sqrt[6]{64a^6b^6} \cdot \sqrt[6]{7ab}$$
$$= -2ab\sqrt[6]{7ab}$$

6. Simplify. Assume that all variables represent positive real numbers.

a. $\sqrt[24]{5^4}$

$$\sqrt[24]{5^4} = (5^4)^{1/24} = 5^{4/24} = 5^{1/6} = \sqrt[6]{5}$$

b. $\sqrt[12]{x^8}$

$$\sqrt[12]{x^8} = (x^8)^{1/12} = x^{8/12} = x^{2/3} = \sqrt[3]{x^2}$$

5. Simplify. Assume that all variables represent positive real numbers.

a. $\sqrt{100y^3}$

_____

b. $\sqrt{48m^5r^9}$

_____

c. $\sqrt[3]{-32n^7t^5}$

_____

d. $-\sqrt[4]{405x^3y^9}$

_____

6. Simplify. Assume that all variables represent positive real numbers.

a. $\sqrt[12]{11^9}$

_____

b. $\sqrt[30]{z^{24}}$

_____

Objective 3 Practice Exercises

For extra help, see Examples 4–6 on pages 452–454 of your text.

Simplify each radical. Assume that variables represent positive real numbers.

7. $\sqrt[42]{x^{28}}$

7. _____

8. $\sqrt{8x^3 y^6 z^{11}}$ 8. _____

9. $\sqrt[3]{1250 a^5 b^7}$ 9. _____

Objective 4   Simplify products and quotients of radicals with different indexes.

Video Examples

Review this example for Objective 4:

7.   Simplify $\sqrt{3} \cdot \sqrt[5]{6}$.

Because the different indexes, 2 and 5, have least common multiple index of 10, we use rational exponents to write each radical as a tenth root.

$$\sqrt{3} = 3^{1/2} = 3^{5/10} = \sqrt[10]{3^5} = \sqrt[10]{243}$$

$$\sqrt[5]{6} = 6^{1/5} = 6^{2/10} = \sqrt[10]{6^2} = \sqrt[10]{36}$$

$$\sqrt{3} \cdot \sqrt[5]{6} = \sqrt[10]{243} \cdot \sqrt[10]{36}$$

$$= \sqrt[10]{243 \cdot 36}$$

$$= \sqrt[10]{8748}$$

Now Try:

7.   Simplify $\sqrt[3]{3} \cdot \sqrt[6]{7}$.

_____

Objective 4 Practice Exercises

For extra help, see Example 7 on page 454 of your text.

Simplify each radical. Assume that variables represent positive real numbers.

10. $\sqrt{r} \cdot \sqrt[3]{r}$ 10. _____

11. $\sqrt[4]{2} \cdot \sqrt[8]{7}$ 11. _____

12. $\sqrt{3} \cdot \sqrt[5]{64}$ 12. _____

Name: Date:

Instructor: Section:

Objective 5   Use the Pythagorean theorem.

Video Examples

Review this example for Objective 5:

8. Use the Pythagorean theorem to find the length of the unknown side of the triangle.

$$a^2 + b^2 = c^2$$

$$12^2 + b^2 = 25^2$$

$$144 + b^2 = 625$$

$$b^2 = 481$$

$$b = \sqrt{481}$$

The length of the side is $\sqrt{481}$.

Now Try:

8. Use the Pythagorean theorem to find the length of the unknown side of the triangle.

_____

Objective 5 Practice Exercises

For extra help, see Example 8 on page 455 of your text.

Find the unknown length in each right triangle. Simplify the answer if necessary.

13.

13. _____

14.

14. _____

15.

15. _____

Objective 6   Use the distance formula.

Video Examples

Review this example for Objective 6:

9. Find the distance between the points
    (2,–2) and (–6, 1).

    Use the distance formula. Let $(x_1, y_1) = (2, -2)$
    and $(x_2, y_2) = (-6, 1)$.

    $$d = \sqrt{(x_2 - x_1)^2 + (y_2 - y_1)^2}$$
    $$= \sqrt{(-6 - 2)^2 + [1 - (-2)]^2}$$
    $$= \sqrt{(-8)^2 + 3^2}$$
    $$= \sqrt{64 + 9}$$
    $$= \sqrt{73}$$

Now Try:

9. Find the distance between the
    points (–1,–2) and (–4, 3).

    _____

Objective 6 Practice Exercises

For extra help, see Example 9 on page 457 of your text.

Find the distance between each pair of points.

16.   (3, 4) and (–1,–2)   16. _____

17.   (–2,–3) and (–5, 1)   17. _____

18.   (4, 2) and (3,–1)   18. _____

Name:
Date:
Instructor:
Section:

Objective 7 Find an equation of a circle given its center and radius.

Video Examples

Review these examples for Objective 7:

10. Write an equation of the circle with radius 2 and center at (0, 0) and graph.

If the point (x, y) is on the circle, then the distance from (x, y) to the center (0, 0) is 2.

$$\sqrt{(x_2 - x_1)^2 + (y_2 - y_1)^2} = d$$

$$\sqrt{(x - 0)^2 + (y - 0)^2} = 2$$

$$\left(\sqrt{x^2 + y^2}\right)^2 = 2^2$$

$$x^2 + y^2 = 4$$

The equation of this circle is $x^2 + y^2 = 4$.

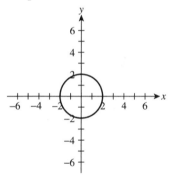

11. Find an equation of the circle with center $(-3, 2)$ and radius 3 and graph it.

$$\sqrt{(x_2 - x_1)^2 + (y_2 - y_1)^2} = d$$

$$\sqrt{(x - (-3))^2 + (y - 2)^2} = 3$$

$$\left(\sqrt{(x + 3)^2 + (y - 2)^2}\right)^2 = 3^2$$

$$(x + 3)^2 + (y - 2)^2 = 9$$

To graph the circle, plot the center $(-3, 2)$, then move three units right, left, up, and down from the center, plotting the points $(0, 2)$, $(-3, 5)$, $(-6, 2)$, and $(-3, -1)$. Draw a smooth curve through the points.

Now Try:

10. Find an equation of the circle with radius 5 and center (0, 0) and graph it.

_____

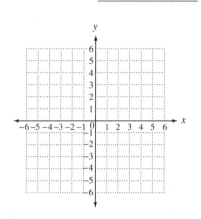

11. Find an equation of the circle with center $(-5, 4)$ and radius 4 and graph it.

_____

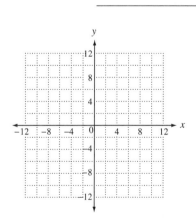

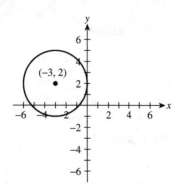

12. Write an equation of a circle with center $(-2, -4)$ and radius $2\sqrt{5}$.

Use the center-radius form.

$$(x-h)^2 + (y-k)^2 = r^2$$

$$[x-(-2)]^2 + [y-(-4)]^2 = \left(2\sqrt{5}\right)^2$$

$$(x+2)^2 + (y+4)^2 = 20$$

12. Write an equation of a circle with center $(3, -1)$ and radius $\sqrt{6}$.

_____

Objective 7 Practice Exercises

For extra help, see Examples 10–12 on pages 457–458 of your text.

Find the equation of a circle satisfying the given conditions.

19.   center: $(3, -4)$; radius: 5

19. _____

20.   center: $(-2, -2)$; radius: 3

20. _____

21.   center: $(0, 3)$; radius: $\sqrt{2}$

21. _____

Chapter 7 ROOTS, RADICALS, AND ROOT FUNCTIONS

7.4   Adding and Subtracting Radical Expressions

Learning Objectives

1   Simplify radical expressions involving addition and subtraction.

Key Terms

Use the vocabulary terms listed below to complete each statement in exercises 1–2.

     like radicals   **unlike radicals**

1.   The expressions $2\sqrt{2}$ and $6\sqrt[3]{2}$ are _____.

2.   The expressions $2\sqrt{2}$ and $7\sqrt{2}$ are _____.

Objective 1   Simplify radical expressions involving addition and subtraction.

Video Examples

Review these examples for Objective 1:

1. Add or subtract to simplify each radical expression.

   a. $3\sqrt{13} + 5\sqrt{52}$

$$3\sqrt{13} + 5\sqrt{52} = 3\sqrt{13} + 5\sqrt{4}\sqrt{13}$$
$$= 3\sqrt{13} + 5 \cdot 2\sqrt{13}$$
$$= 3\sqrt{13} + 10\sqrt{13}$$
$$= (3 + 10)\sqrt{13}$$
$$= 13\sqrt{13}$$

   b. $\sqrt{48x} - \sqrt{12x}, \ x \geq 0$

$$\sqrt{48x} - \sqrt{12x} = \sqrt{16} \cdot \sqrt{3x} - \sqrt{4} \cdot \sqrt{3x}$$
$$= 4\sqrt{3x} - 2\sqrt{3x}$$
$$= (4 - 2)\sqrt{3x}$$
$$= 2\sqrt{3x}$$

   c. $7\sqrt{3} - 6\sqrt{21}$

The radicands differ and are already simplified, so this expression cannot be simplified further.

Now Try:

1. Add or subtract to simplify each radical expression.

   a. $3\sqrt{54} - 5\sqrt{24}$

   _____

   b. $3\sqrt{18z} + 2\sqrt{8z}, \ z \geq 0$

   _____

   c. $3\sqrt{7} + 2\sqrt{6}$

   _____

Name: _____ Date: _____

Instructor: _____ Section: _____

2. Simplify. Assume that all variables represent positive real numbers.

a. $7\sqrt[4]{32} - 9\sqrt[4]{2}$

$$7\sqrt[4]{32} - 9\sqrt[4]{2} = 7\sqrt[4]{16}\cdot\sqrt[4]{2} - 9\sqrt[4]{2}$$
$$= 7\cdot 2\cdot\sqrt[4]{2} - 9\sqrt[4]{2}$$
$$= 14\sqrt[4]{2} - 9\sqrt[4]{2}$$
$$= (14 - 9)\sqrt[4]{2}$$
$$= 5\sqrt[4]{2}$$

b. $6\sqrt[3]{27x^5 r} + 2x\sqrt[3]{x^2 r}$

$$6\sqrt[3]{27x^5 r} + 2x\sqrt[3]{x^2 r}$$
$$= 6\cdot\sqrt[3]{27x^3}\cdot\sqrt[3]{x^2 r} + 2x\sqrt[3]{x^2 r}$$
$$= 18x\sqrt[3]{x^2 r} + 2x\sqrt[3]{x^2 r}$$
$$= (18x + 2x)\sqrt[3]{x^2 r}$$
$$= 20x\sqrt[3]{x^2 r}$$

c. $3\sqrt{40x^5} + 5\sqrt[3]{48x^5}$

$$3\sqrt{40x^5} + 5\sqrt[3]{48x^5}$$
$$= 3\cdot\sqrt{4x^4\cdot 10x} + 5\sqrt[3]{8x^3\cdot 6x^2}$$
$$= 3\cdot\sqrt{4x^4}\cdot\sqrt{10x} + 5\sqrt[3]{8x^3}\cdot\sqrt[3]{6x^2}$$
$$= 3\cdot 2x^2\cdot\sqrt{10x} + 5\cdot 2x\cdot\sqrt[3]{6x^2}$$
$$= 6x^2\sqrt{10x} + 10x\sqrt[3]{6x^2}$$

2. Simplify. Assume that all variables represent positive real numbers.

a. $7\sqrt[3]{54} - 6\sqrt[3]{128}$

_____

b. $\sqrt[4]{32y^2 z^5} + 3z\sqrt[4]{2y^2 z}$

_____

c. $2\sqrt[3]{54x^7} + 2\sqrt{27x^7}$

_____

3. Simplify. Assume that all variables represent positive real numbers.

a. $\dfrac{\sqrt{32}}{3} + \dfrac{\sqrt{8}}{\sqrt{18}}$

$$\dfrac{\sqrt{32}}{3} + \dfrac{\sqrt{8}}{\sqrt{18}} = \dfrac{\sqrt{16 \cdot 2}}{3} + \dfrac{\sqrt{4 \cdot 2}}{\sqrt{9 \cdot 2}}$$

$$= \dfrac{4\sqrt{2}}{3} + \dfrac{2\sqrt{2}}{3\sqrt{2}}$$

$$= \dfrac{4\sqrt{2}}{3} + \dfrac{2}{3}$$

$$= \dfrac{4\sqrt{2} + 2}{3}$$

b. $\sqrt[3]{\dfrac{81}{y^6}} + 5\sqrt[3]{\dfrac{27}{y^3}}$

$$\sqrt[3]{\dfrac{81}{y^6}} + 5\sqrt[3]{\dfrac{27}{y^3}} = \dfrac{\sqrt[3]{81}}{\sqrt[3]{y^6}} + 5\dfrac{\sqrt[3]{27}}{\sqrt[3]{y^3}}$$

$$= \dfrac{3\sqrt[3]{3}}{y^2} + 5\left(\dfrac{3}{y}\right)$$

$$= \dfrac{3\sqrt[3]{3}}{y^2} + \dfrac{15}{y}$$

$$= \dfrac{3\sqrt[3]{3} + 15y}{y^2}$$

3. Simplify. Assume that all variables represent positive real numbers.

a. $\sqrt{\dfrac{10}{18}} + \dfrac{\sqrt{15}}{\sqrt{27}}$

_____

b. $\sqrt[3]{\dfrac{216}{w^6}} + \sqrt{\dfrac{121}{w^4}}$

_____

Objective 1 Practice Exercises

For extra help, see Examples 1–3 on pages 463–465 of your text.

Add or subtract. Assume that all variables represent positive real numbers.

1. $\sqrt{100x} - \sqrt{9x} + \sqrt{25x}$

1. _____

2. $2\sqrt[3]{16r} + \sqrt[3]{54r} - \sqrt[3]{16r}$

2. _____

3. $\sqrt[3]{\dfrac{y^7}{125}} + y^2\sqrt[3]{\dfrac{y}{27}}$

3. _____

Chapter 7 ROOTS, RADICALS, AND ROOT FUNCTIONS

7.5   Multiplying and Dividing Radical Expressions

Learning Objectives
1   Multiply radical expressions.
2   Rationalize denominators with one radical term.
3   Rationalize denominators with binomials involving radicals.
4   Write radical quotients in lowest terms.

Key Terms

Use the vocabulary terms listed below to complete each statement in exercises 1−2.

  rationalizing the denominator   **conjugate**

1.   The _____ of $a + b$ is $a - b$.

2.   The process of removing radicals from the denominator so that the denominator contains only rational quantities is called _____.

Objective 1   Multiply radical expressions.

Video Examples

Review these examples for Objective 1:
2.   Multiply, using the FOIL method.

  a. $\left(\sqrt{6}+\sqrt{5}\right)\left(\sqrt{6}-\sqrt{5}\right)$

  This is the difference of squares.
  $$\left(\sqrt{6}+\sqrt{5}\right)\left(\sqrt{6}-\sqrt{5}\right)=\left(\sqrt{6}\right)^2-\left(\sqrt{5}\right)^2$$
  $$=6-5$$
  $$=1$$

  b. $\left(5-\sqrt{3}\right)\left(\sqrt{2}+\sqrt{5}\right)$

  $\left(5-\sqrt{3}\right)\left(\sqrt{2}+\sqrt{5}\right)$
  $=5\cdot\sqrt{2}+5\cdot\sqrt{5}-\sqrt{3}\cdot\sqrt{2}-\sqrt{3}\cdot\sqrt{5}$
  $=5\sqrt{2}+5\sqrt{5}-\sqrt{6}-\sqrt{15}$

Now Try:
2.   Multiply, using the FOIL method.

  a. $\left(\sqrt{14}-\sqrt{2}\right)\left(\sqrt{14}+\sqrt{2}\right)$

  _____

  b. $\left(3-\sqrt{2}\right)\left(2+\sqrt{7}\right)$

  _____

c. $\left(\sqrt{11}-6\right)^2$

$$\begin{aligned}\left(\sqrt{11}-6\right)^2 &=\left(\sqrt{11}-6\right)\left(\sqrt{11}-6\right)\\&=\sqrt{11}\cdot\sqrt{11}-6\sqrt{11}-6\sqrt{11}+6\cdot 6\\&=11-12\sqrt{11}+36\\&=47-12\sqrt{11}\end{aligned}$$

c. $\left(3-\sqrt{2}\right)^2$

_____

Objective 1 Practice Exercises

For extra help, see Examples 1–2 on pages 468–469 of your text.

Multiply each product, then simplify. Assume that variables represent positive real numbers.

1. $\left(\sqrt{5}+\sqrt{6}\right)\left(\sqrt{2}-4\right)$

1. _____

2. $\left(\sqrt{2}-\sqrt{12}\right)^2$

2. _____

3. $\left(2+\sqrt[3]{5}\right)\left(2-\sqrt[3]{5}\right)$

3. _____

Objective 2   Rationalize denominators with one radical term.

Video Examples

Review these examples for Objective 2:

3. Rationalize the denominator.

$$\frac{4}{\sqrt{3}}$$

$$\frac{4}{\sqrt{3}} = \frac{4 \cdot \sqrt{3}}{\sqrt{3} \cdot \sqrt{3}} = \frac{4\sqrt{3}}{3}$$

4. Simplify the radical.

$$-\sqrt{\frac{27}{98}}$$

$$-\sqrt{\frac{27}{98}} = -\frac{\sqrt{27}}{\sqrt{98}}$$

$$= -\frac{\sqrt{9 \cdot 3}}{\sqrt{49 \cdot 2}}$$

$$= -\frac{3\sqrt{3}}{7\sqrt{2}}$$

$$= -\frac{3\sqrt{3} \cdot \sqrt{2}}{7\sqrt{2} \cdot \sqrt{2}}$$

$$= -\frac{3\sqrt{6}}{7 \cdot 2}$$

$$= -\frac{3\sqrt{6}}{14}$$

5. Simplify.

$$\sqrt[3]{\frac{16}{9}}$$

$$\sqrt[3]{\frac{16}{9}} = \frac{\sqrt[3]{8 \cdot 2}}{\sqrt[3]{9}} = \frac{2\sqrt[3]{2}}{\sqrt[3]{9}}$$

$$= \frac{2\sqrt[3]{2} \cdot \sqrt[3]{3}}{\sqrt[3]{9} \cdot \sqrt[3]{3}}$$

$$= \frac{2\sqrt[3]{6}}{\sqrt[3]{27}}$$

$$= \frac{2\sqrt[3]{6}}{3}$$

Now Try:

3. Rationalize the denominator.

$$\frac{2}{\sqrt{15}}$$

_____

4. Simplify the radical.

$$-\sqrt{\frac{45}{32}}$$

_____

5. Simplify.

$$\sqrt[3]{\frac{8}{100}}$$

_____

Objective 2 Practice Exercises

For extra help, see Examples 3–5 on pages 470–472 of your text.

Simplify. Assume that variables represent positive real numbers.

4. $\sqrt{\dfrac{5a^2b^3}{6}}$

4. _____

5. $\sqrt{\dfrac{7y^2}{12b}}$

5. _____

6. $\sqrt[3]{\dfrac{5}{49x}}$

6. _____

Objective 3   Rationalize denominators with binomials involving radicals.

Video Examples

Review this example for Objective 3:

6. Rationalize the denominator.

$$\frac{5}{5+\sqrt{2}}$$

$$\frac{5}{5+\sqrt{2}} = \frac{5(5-\sqrt{2})}{(5+\sqrt{2})(5-\sqrt{2})}$$

$$= \frac{5(5-\sqrt{2})}{25-2}$$

$$= \frac{5(5-\sqrt{2})}{23}$$

Now Try:

6. Rationalize the denominator.

$$\frac{2}{\sqrt{3}-2}$$

_____

Objective 3 Practice Exercises

For extra help, see Example 6 on page 473 of your text.

Rationalize each denominator. Write quotients in lowest terms. Assume that variables represent positive real numbers.

7. $\dfrac{4}{\sqrt{3}+2}$

7. _____

8. $\dfrac{5}{\sqrt{3}-\sqrt{10}}$

8. _____

9. $\dfrac{\sqrt{6}+2}{\sqrt{2}-4}$

9. _____

Objective 4   Write radical quotients in lowest terms.

Video Examples

Review this example for Objective 4:

7.   Write the quotient in lowest terms.

$$\frac{72\sqrt{2}-16\sqrt{7}}{24}$$

$$\frac{72\sqrt{2}-16\sqrt{7}}{24}=\frac{8\left(9\sqrt{2}-2\sqrt{7}\right)}{24}$$

$$=\frac{9\sqrt{2}-2\sqrt{7}}{3}$$

Now Try:

7.   Write the quotient in lowest terms.

$$\frac{9+6\sqrt{15}}{12}$$

_____

Objective 4 Practice Exercises

For extra help, see Example 7 on page 474 of your text.

Write each quotient in lowest terms. Assume that variables represent positive real numbers.

10. $\dfrac{7-\sqrt{98}}{14}$

10. _____

11. $\dfrac{16-12\sqrt{72}}{24}$

11. _____

12. $\dfrac{2x-\sqrt{8x^2}}{4x}$

12. _____

Chapter 7 ROOTS, RADICALS, AND ROOT FUNCTIONS

7.6 Solving Equations with Radicals

Learning Objectives
1 Solve radical equations using the power rule.
2 Solve radical equations with indexes greater than 2.
3 Use the power rule to solve a formula for a specified variable.

Key Terms

Use the vocabulary terms listed below to complete each statement in exercises 1–2.

radical equation **extraneous solution**

1. A(n) _____ is a potential solution to an equation that does not satisfy the equation.

2. An equation with a variable in the radicand is a(n) _____.

Objective 1 Solve radical equations using the power rule.

Video Examples

Review these examples for Objective 1:

1. Solve $\sqrt{3w+4} = 7$.

$$\left(\sqrt{3w+4}\right)^2 = 7^2$$
$$3w+4 = 49$$
$$3w = 45$$
$$w = 15$$

Check $\quad \sqrt{3w+4} = 7$

$$\sqrt{3(15)+4} \overset{?}{=} 7$$
$$\sqrt{49} \overset{?}{=} 7$$
$$7 = 7 \quad \text{True}$$

Since 15 satisfies the original equation, the solution set is $\{15\}$.

2. Solve $\sqrt{12p+1}+7 = 0$.

Step 1 $\quad \sqrt{12p+1} = -7$

Step 2 $\left(\sqrt{12p+1}\right)^2 = (-7)^2$

Now Try:

1. Solve $\sqrt{7x-6} = 8$.

_____

2. Solve $\sqrt{4x-19}+5 = 0$.

_____

Step 3 $12p + 1 = 49$

$12p = 48$

$p = 4$

Step 4 Check $\sqrt{12p+1} + 7 = 0$

$\sqrt{12(4)+1} + 7 \overset{?}{=} 0$

$\sqrt{49} + 7 \overset{?}{=} 0$

$14 = 0$ False

The false result shows that the proposed solution 4 is not a solution of the original equation. It is extraneous. The solution set is $\varnothing$.

3.   Solve $\sqrt{x+3} = x - 3$.

Step 1 The radical is isolated on the left side of the equation.

Step 2 Square each side.

$\left(\sqrt{x+3}\right)^2 = (x-3)^2$

$x + 3 = x^2 - 6x + 9$

Step 3 Write the equation in standard form and solve.

$0 = x^2 - 7x + 6$

$0 = (x-1)(x-6)$

$x - 1 = 0$ or $x - 6 = 0$

$x = 1$ or $x = 6$

Step 4 Check each proposed solution in the original equation.

$\sqrt{x+3} = x - 3$ $\sqrt{x+3} = x - 3$

$\sqrt{1+3} \overset{?}{=} 1 - 3$ $\sqrt{6+3} \overset{?}{=} 6 - 3$

$\sqrt{4} \overset{?}{=} -2$ $\sqrt{9} \overset{?}{=} 3$

$2 = -2$ False $3 = 3$ True

The solution set is $\{6\}$. The other proposed solution, 1, is extraneous.

3.   Solve $\sqrt{x+11} = x - 1$.

_____

5. Solve $\sqrt{3x} - 4 = \sqrt{x-2}$.

$$\left(\sqrt{3x} - 4\right)^2 = \left(\sqrt{x-2}\right)^2$$
$$3x - 8\sqrt{3x} + 16 = x - 2$$
$$-8\sqrt{3x} = -2x - 18$$
$$\left(-8\sqrt{3x}\right)^2 = (-2x - 18)^2$$
$$192x = 4x^2 + 72x + 324$$
$$0 = 4x^2 - 120x + 324$$
$$0 = 4\left(x^2 - 30x + 81\right)$$
$$0 = 4(x-3)(x-27)$$
$$x - 3 = 0 \quad \text{or} \quad x - 27 = 0$$
$$x = 3 \quad \text{or} \quad x = 27$$

Check

$$\sqrt{3x} - 4 = \sqrt{x-2} \qquad \sqrt{3x} - 4 = \sqrt{x-2}$$
$$\sqrt{3(3)} - 4 \overset{?}{=} \sqrt{3-2} \qquad \sqrt{3(27)} - 4 \overset{?}{=} \sqrt{27-2}$$
$$3 - 4 \overset{?}{=} \sqrt{1} \qquad 9 - 4 \overset{?}{=} \sqrt{25}$$
$$-1 = 1 \quad \text{False} \qquad 5 = 5 \quad \text{True}$$

The proposed solution, 27, is valid, but 3 is extraneous and must be rejected. The solution set is {27}.

5. Solve $\sqrt{3x+4} = \sqrt{9x} - 2$.

_____

Objective 1 Practice Exercises

For extra help, see Examples 1–5 on pages 479–482 of your text.

Solve each equation.

1. $\sqrt{4x - 19} = 5$

1. _____

2. $\sqrt{12p + 1} + 7 = 0$

2. _____

3. $\sqrt{k+10} + \sqrt{2k+19} = 2$

3. _____

Objective 2   Solve radical equations with indexes greater than 2.

Video Examples

Review this example for Objective 2:

6. Solve $\sqrt[3]{5r-6} = \sqrt[3]{3r+4}$.

$$\left(\sqrt[3]{5r-6}\right)^3 = \left(\sqrt[3]{3r+4}\right)^3$$

$$5r - 6 = 3r + 4$$

$$2r = 10$$

$$r = 5$$

Check $\sqrt[3]{5r-6} = \sqrt[3]{3r+4}$

$$\sqrt[3]{5(5)-6} \overset{?}{=} \sqrt[3]{3(5)+4}$$

$$\sqrt[3]{19} = \sqrt[3]{19}   \text{True}$$

The solution set is $\{5\}$.

Now Try:

6. Solve $\sqrt[4]{8x+5} = \sqrt[4]{7x+7}$.

_____

Objective 2 Practice Exercises

For extra help, see Example 6 on pages 482–483 of your text.

Solve each equation.

4. $\sqrt[3]{2a-63} + 5 = 0$

4. _____

5. $\sqrt[5]{5a+1} - \sqrt[5]{2a-11} = 0$

5. _____

6. $\sqrt[4]{8x+5} = \sqrt[4]{7x+7}$

6. _____

Objective 3   Use the power rule to solve a formula for a specified variable.

Video Examples

Review this example for Objective 3:

7. Solve the formula $d = \sqrt{\dfrac{H}{1.6n}}$ for n.

$$d^2 = \left(\sqrt{\dfrac{H}{1.6n}}\right)^2$$

$$d^2 = \dfrac{H}{1.6n}$$

$$1.6d^2 n = H$$

$$n = \dfrac{H}{1.6d^2}$$

Now Try:

7. Solve the formula $r = \sqrt{\dfrac{3v}{\pi h}}$ for h.

_____

Objective 3 Practice Exercises

For extra help, see Example 7 on page 483 of your text.

Solve each equation for the indicated variable.

7. $Z = \sqrt{\dfrac{L}{C}}$, for L

7. _____

8. $f = \dfrac{1}{2\pi\sqrt{LC}}$, for C

8. _____

9. $N = \dfrac{1}{2\pi}\sqrt{\dfrac{a}{r}}$, for r

9. _____

Chapter 7 ROOTS, RADICALS, AND ROOT FUNCTIONS

7.7   Complex Numbers

Learning Objectives

1   Simplify numbers of the form $\sqrt{-b}$, where $b > 0$.
2   Recognize subsets of the complex numbers.
3   Add and subtract complex numbers.
4   Multiply complex numbers.
5   Divide complex numbers.
6   Simplify powers of i.

Key Terms

Use the vocabulary terms listed below to complete each statement in exercises 1–7.

complex number   real part   imaginary part

pure imaginary number   standard form (of a complex number)

nonreal complex number   complex conjugate

1.   A _____ is a number that can be written in the form $a + bi$, where a and b are real numbers.

2.   The _____ of $a + bi$ is $a - bi$.

3.   The _____ of $a + bi$ is bi.

4.   The _____ of $a + bi$ is a.

5.   A complex number is in _____ if it is written in the form $a + bi$.

6.   A complex number $a + bi$ with $a = 0$ and $b \neq 0$ is called a
_____.

7.   A complex number $a + bi$ with $b \neq 0$ is called a _____.

Objective 1   Simplify numbers of the form $\sqrt{-b}$, where $b > 0$.

Video Examples

Review these examples for Objective 1:
1.   Write the number as a product of a real number and i.

$$\sqrt{-36}$$

$$\sqrt{-36} = i\sqrt{36} = 6i$$

Now Try:
1.   Write the number as a product of a real number and i.

$$\sqrt{-16}$$

_____

2. Multiply.

$$\sqrt{-6}\cdot\sqrt{-7}$$

$$\sqrt{-6}\cdot\sqrt{-7}=i\sqrt{6}\cdot i\sqrt{7}$$

$$=i^2\sqrt{6\cdot 7}$$

$$=(-1)\sqrt{42}$$

$$=-\sqrt{42}$$

3. Divide.

$$\frac{\sqrt{-125}}{\sqrt{-5}}$$

$$\frac{\sqrt{-125}}{\sqrt{-5}}=\frac{i\sqrt{125}}{i\sqrt{5}}$$

$$=\sqrt{\frac{125}{5}}$$

$$=\sqrt{25}$$

$$=5$$

2. Multiply.

$$\sqrt{-5}\cdot\sqrt{-6}$$

_____

3. Divide.

$$\frac{\sqrt{-200}}{\sqrt{-8}}$$

_____

Objective 1 Practice Exercises

For extra help, see Examples 1–3 on pages 486–487 of your text.

Write the number as a product of a real number and i. Simplify all radical expressions.

1. $-\sqrt{-162}$

1. _____

Multiply or divide as indicated

2. $\sqrt{-5}\cdot\sqrt{-3}\cdot\sqrt{-7}$

2. _____

3. $\dfrac{\sqrt{-42}\cdot\sqrt{-6}}{\sqrt{-7}}$

3. _____

Objective 2   Recognize subsets of the complex numbers.

For extra help, see page 487 of your text.

Classify each of the following complex numbers as real *or* imaginary.

4. $\sqrt{5}$ 4. _____

5. $\sqrt{3} - i\sqrt{5}$ 5. _____

6. $i\sqrt{7}$ 6. _____

Objective 3   Add and subtract complex numbers.

Video Examples

Review these examples for Objective 3: | **Now Try:**
4.   Add.

$(2 + 9i) + (10 - 3i)$

$(2 + 9i) + (10 - 3i) = (2 + 10) + (9 - 3)i$
$\qquad = 12 + 6i$

5.   Subtract.

$(7 - 9i) - (-5 - 6i)$

$(7 - 9i) - (-5 - 6i) = [7 - (-5)] + [-9 - (-6)]i$
$\qquad = (7 + 5) + (-9 + 6)i$
$\qquad = 12 - 3i$

Now Try:
4.   Add.

$(4 - 7i) + (6 - 2i)$

_____

5.   Subtract.

$(12 + 2i) - (-12 - 2i)$

_____

Objective 3 Practice Exercises

For extra help, see Examples 4–5 on page 488 of your text.

Add or subtract as indicated. Write answers in standard form.

7. $(-7 - 2i) - (-3 - 3i)$ 7. _____

8. $4i - (9 + 5i) + (2 + 3i)$ 8. _____

9. $(7 - 9i) - (5 - 6i)$ 9. _____

Objective 4   Multiply complex numbers.

Video Examples

Review this example for Objective 4:

6. Multiply.

$$6i(2-7i)$$

$$6i(2-7i) = 6i(2) + 6i(-7i)$$
$$= 12i - 42i^2$$
$$= 12i - 42(-1)$$
$$= 42 + 12i$$

Now Try:

6. Multiply.

$$2i(4+7i)$$

_____

Objective 4 Practice Exercises

For extra help, see Example 6 on pages 488–489 of your text.

Multiply.

10. $\quad (2-5i)(2+5i)$ **10.** _____

11. $\quad (1+3i)^2$ **11.** _____

12. $\quad (12+2i)(-1+i)$ **12.** _____

Objective 5   Divide complex numbers.

Video Examples

Review this example for Objective 5:

7. Find the quotient.

$$\frac{6-i}{2-3i}$$

Multiply the numerator and denominator by $2 + 3i$, the conjugate of the denominator.

$$\frac{6-i}{2-3i} = \frac{(6-i)(2+3i)}{(2-3i)(2+3i)}$$

$$= \frac{12+18i-2i-3i^2}{2^2+3^2}$$

$$= \frac{12+16i-3(-1)}{4+9}$$

$$= \frac{15+16i}{13}, \text{ or } \frac{15}{13}+\frac{16}{13}i$$

Now Try:

7. Find the quotient.

$$\frac{4+i}{5-2i}$$

_____

Objective 5 Practice Exercises

For extra help, see Example 7 on pages 489–490 of your text.

Write each quotient in the form a + bi.

13. $\frac{3-2i}{2+i}$

13. _____

14. $\frac{5+2i}{9-4i}$

14. _____

15. $\frac{6-i}{2-3i}$

15. _____

Objective 6   Simplify powers of *i*.

Video Examples

Review this example for Objective 6:
8.   Find the power of *i*.

i^{100}

$$i^{100} = (i^4)^{25} = 1^{25} = 1$$

Now Try:
8.   Find the power of *i*.

i^{48}

_____

Objective 6 Practice Exercises

For extra help, see Example 8 on page 490 of your text.

Find each power of i.

16. i^{14}

16. _____

17. i^{113}

17. _____

18. i^{-21}

18. _____

Chapter 8   QUADRATIC EQUATIONS, INEQUALITIES, AND FUNCTIONS

8.1   The Square Root Property and Completing the Square

Learning Objectives
1   Review the zero-factor property.
2   Learn the square root property.
3   Solve quadratic equations of the form $(ax + b)^2 = c$ by extending the square root property.
4   Solve quadratic equations by completing the square.
5   Solve quadratic equations with solutions that are not real numbers.

Key Terms

Use the vocabulary terms listed below to complete each statement in exercises 1−4.

> **quadratic equation**   **zero-factor property**
>
> **completing the square**   **square root property**

1. The _____ says that, if k is positive and $a^2 = k$, then $a = \pm\sqrt{k}$.

2. Use the process called _____ in order to rewrite an equation so it can be solved using the square root property.

3. An equation that can be written in the form $ax^2 + bx + c = 0$ is a
    _____.

4. The _____ states that if a product equals 0, then at least one of the factors of the product also equals zero.

Objective 1   Review the zero-factor property.

Objective 1 Practice Exercises

For extra help, see Example 1 on page 504 of your text.

Solve each equation by factoring.

1. $15s^2 - 2 = s$ 1. _____

2. $z^2 = 6z - 9$ 2. _____

3. $16m^2 - 64 = 0$ **3.** _____

Objective 2   Learn the square root property.

Video Examples

Review these examples for Objective 2:

2.   Solve the equation.

$x^2 = 7$

$$x^2 = 7$$
$$x = \sqrt{7} \text{ or } x = -\sqrt{7}$$

The solution set is $\{\sqrt{7}, -\sqrt{7}\}$, or $\{\pm\sqrt{7}\}$.

3.   Use Galileo's formula to determine how long it will take a penny dropped from the 86th floor Observatory deck of the Empire State Building to reach the ground. The deck is 1050 feet above the ground. Round your answer to the nearest tenth.

Galileo's formula is $d = 16t^2$, where d is the distance in feet that an object falls, and t is the time in seconds.

$$d = 16t^2$$

$$1050 = 16t^2$$

$$65.625 = t^2$$
$$t = \sqrt{65.625} \text{ or } t = -\sqrt{65.625}$$

Time cannot be negative, so we discard $t = -\sqrt{65.625}$. Using a calculator, $\sqrt{65.625} \approx 8.1$, so $t \approx 8.1$. The penny would fall to the ground in about 8.1 seconds.

Now Try:

2.   Solve the equation.

$r^2 = 13$

_____

3.   A child dropped a ball from a hotel balcony that is 113 ft above the ground. Use Galileo's formula to determine how long it takes for the ball to reach the ground. Round your answer to the nearest tenth.

_____

Objective 2 Practice Exercises

For extra help, see Examples 2–3 on pages 505–506 of your text.

Solve each equation by using the square root property. Express all radicals in simplest form.

4. $r^2 = 30$ **4.** _____

5. $x^2 - 98 = 0$

5. _____

6. $3d^2 - 750 = 0$

6. _____

Objective 3 **Solve equations of the form** $(ax+b)^2 = k$ **by extending the square root property.**

Video Examples

Review this example for Objective 3: | **Now Try:**

5. Solve $(3x+4)^2 = 40$. | **5.** Solve $(2x+5)^2 = 32$.

$$(3x+4)^2 = 40$$

$$3x+4 = \sqrt{40} \qquad \text{or} \quad 3x+4 = -\sqrt{40}$$

$$3x = -4 + \sqrt{40} \quad \text{or} \qquad 3x = -4 - \sqrt{40}$$

$$x = \frac{-4+2\sqrt{10}}{3} \quad \text{or} \qquad x = \frac{-4-2\sqrt{10}}{3}$$

_____

Check

$$(3x+4)^2 = 40$$

$$\left[3\left(\frac{-4-2\sqrt{10}}{3}\right)+4\right]^2 \overset{?}{=} 40$$

$$\left(-4-2\sqrt{10}+4\right)^2 \overset{?}{=} 40$$

$$\left(-2\sqrt{10}\right)^2 \overset{?}{=} 40$$

$$40 = 40$$

The check for the second solution is similar.

The solution sets is $\left\{\dfrac{-4-2\sqrt{10}}{3}, \dfrac{-4+2\sqrt{10}}{3}\right\}$.

Objective 3 Practice Exercises

For extra help, see Examples 4–5 on pages 506–507 of your text.

Solve each equation by using the square root property. Express all radicals in simplest form.

7. $(y+2)^2 = 16$

7. _____

8. $(q-4)^2 = 7$

8. _____

9. $(3f+4)^2 = 32$

9. _____

Objective 4   Solve quadratic equations by completing the square.

Video Examples

Review these examples for Objective 4:

6.   Solve $x^2 - 12x + 24 = 0$.

$$x^2 - 12x + 24 = 0$$
$$x^2 - 12x = -24$$

Take half of the coefficient of the first-degree term, $-12x$, and square the result.

$$\left[\frac{1}{2}(-12)\right]^2 = (-6)^2 = 36$$

Add 36 to each side.

$$x^2 - 12x + 36 = -24 + 36$$
$$(x-6)^2 = 12$$
$$x - 6 = \sqrt{12} \qquad \text{or} \quad x - 6 = -\sqrt{12}$$
$$x = 6 + 2\sqrt{3} \quad \text{or} \qquad x = 6 - 2\sqrt{3}$$

A check indicates the solution set is

$$\left\{6 - 2\sqrt{3},\ 6 + 2\sqrt{3}\right\}.$$

Now Try:

6.   Solve $x^2 - 6x + 1 = 0$.

_____

7. Solve $x^2 + 5x + 2 = 0$.

Since the coefficient of the second-degree term is 1, begin with Step 2.

Step 2 $x^2 + 5x = -2$

Step 3 Take half the coefficient of the first-degree term and square the result.

$$\left[\frac{1}{2}(5)\right]^2 = \left(\frac{5}{2}\right)^2 = \frac{25}{4}$$

$$x^2 + 5x + \frac{25}{4} = -2 + \frac{25}{4}$$

$$\left(x + \frac{5}{2}\right)^2 = \frac{17}{4}$$

Step 4

$$x + \frac{5}{2} = \sqrt{\frac{17}{4}} \qquad \text{or} \quad x + \frac{5}{2} = -\sqrt{\frac{17}{4}}$$

$$x + \frac{5}{2} = \frac{\sqrt{17}}{2} \qquad \text{or} \quad x + \frac{5}{2} = -\frac{\sqrt{17}}{2}$$

$$x = -\frac{5}{2} + \frac{\sqrt{17}}{2} \quad \text{or} \qquad x = -\frac{5}{2} - \frac{\sqrt{17}}{2}$$

A check shows the solution set is

$$\left\{-\frac{5}{2} - \frac{\sqrt{17}}{2}, -\frac{5}{2} + \frac{\sqrt{17}}{2}\right\}.$$

8. Solve $3x^2 - 6x - 2 = 0$.

$$x^2 - 2x - \frac{2}{3} = 0 \qquad \text{Step 1}$$

$$x^2 - 2x = \frac{2}{3} \qquad \text{Step 2}$$

$$\left[\frac{1}{2}(-2)\right]^2 = (-1)^2 = 1 \qquad \text{Step 3}$$

$$x^2 - 2x + 1 = \frac{2}{3} + 1$$

$$(x - 1)^2 = \frac{5}{3}$$

7. Solve $x^2 - 11x + 8 = 0$.

———————

8. Solve $2p^2 + 6p - 1 = 0$.

———————

$$x - 1 = \sqrt{\frac{5}{3}} \quad \text{or} \quad x - 1 = -\sqrt{\frac{5}{3}} \qquad \text{Step 4}$$

$$x = 1 + \sqrt{\frac{5}{3}} \quad \text{or} \quad x = 1 - \sqrt{\frac{5}{3}}$$

$$x = 1 + \frac{\sqrt{15}}{3} \quad \text{or} \quad x = 1 - \frac{\sqrt{15}}{3}$$

$$x = \frac{3 + \sqrt{15}}{3} \quad \text{or} \quad x = \frac{3 - \sqrt{15}}{3}$$

The solution set is $\left\{ \dfrac{3 - \sqrt{15}}{3}, \ \dfrac{3 + \sqrt{15}}{3} \right\}$.

Objective 4 Practice Exercises

For extra help, see Examples 6–8 on pages 508–510 of your text.

Solve each equation by completing the square.

10. $x^2 - 4x = 2$ **10.** _____

11. $x^2 - 9x + 8 = 0$ **11.** _____

12. $6q^2 + 4q = 1$ **12.** _____

Objective 5   Solve quadratic equations with solutions that are not real numbers.

Video Examples

Review this example for Objective 5:

9.   Solve the equation.

$$x^2 = -48$$

$$x^2 = -48$$
$$x = \sqrt{-48} \quad \text{or} \quad x = -\sqrt{-48}$$
$$x = 4i\sqrt{3} \quad \text{or} \quad x = -4i\sqrt{3}$$

The solution set is $\left\{ -4i\sqrt{3}, \ 4i\sqrt{3} \right\}$.

Now Try:

9.   Solve the equation.

$$y^2 = -32$$

_____

Objective 5 Practice Exercises

For extra help, see Example 9 on page 510 of your text.

Find the complex solutions of each equation.

13. $(10m - 5)^2 + 9 = 0$

13. _____

14. $(m + 1)^2 = -36$

14. _____

15. $(x - 1)^2 + 2 = 0$

15. _____

Chapter 8   QUADRATIC EQUATIONS, INEQUALITIES, AND FUNCTIONS

8.2   The Quadratic Formula

Learning Objectives
1   Derive the quadratic formula.
2   Solve quadratic equations by using the quadratic formula.
3   Use the discriminant to determine the number and type of solutions.

Key Terms

Use the vocabulary terms listed below to complete each statement in exercises 1–2.

     quadratic formula   **discriminant**

1. The expression under the radical in the quadratic formula is called the

            _____ .

2. The formula $x = \dfrac{-b \pm \sqrt{b^2 - 4ac}}{2a}$ is called the _____ .

Objective 1   Derive the quadratic formula.

For extra help, see pages 513–514 of your text.

Objective 2   Solve quadratic equations by using the quadratic formula.

Video Examples

Review these examples for Objective 2:

1. Solve $5x^2 - 13x - 6 = 0$.

    Use the quadratic formula with $a = 5$, $b = -13$, and $c = -6$.

$$x = \frac{-b \pm \sqrt{b^2 - 4ac}}{2a}$$

$$x = \frac{-(-13) \pm \sqrt{(-13)^2 - 4(5)(-6)}}{2(5)}$$

$$x = \frac{13 \pm \sqrt{169 + 120}}{10}$$

$$x = \frac{13 \pm \sqrt{289}}{10}$$

$$x = \frac{13 \pm 17}{10}$$

    There are two solutions.

Now Try:

1. Solve $6x^2 - 17x + 12 = 0$.

                 _____

$$x = \frac{13 + 17}{10} = 3 \quad \text{or} \quad x = \frac{13 - 17}{10} = \frac{-4}{10} = -\frac{2}{5}$$

The solution set is $\left\{ -\frac{2}{5},\ 3 \right\}$.

2. Solve $4x^2 = -4x + 1$.

First write the equation in standard

form as $4x^2 + 4x - 1 = 0$.

$a = 4,\ b = 4,\ c = -1$

$$x = \frac{-b \pm \sqrt{b^2 - 4ac}}{2a}$$

$$x = \frac{-4 \pm \sqrt{(4)^2 - 4(4)(-1)}}{2(4)}$$

$$x = \frac{-4 \pm \sqrt{16 + 16}}{8} = \frac{-4 \pm \sqrt{32}}{8}$$

$$x = \frac{-4 \pm 4\sqrt{2}}{8}$$

$$x = \frac{4(-1 \pm \sqrt{2})}{4(2)}$$

$$x = \frac{-1 \pm \sqrt{2}}{2}$$

The solution set is $\left\{ \dfrac{-1 - \sqrt{2}}{2},\ \dfrac{-1 + \sqrt{2}}{2} \right\}$.

3. Solve $(5x - 2)(x + 2) = -9$.

$$(5x - 2)(x + 2) = -9$$
$$5x^2 + 8x - 4 = -9$$
$$5x^2 + 8x + 5 = 0$$

From the standard form, we identify

$a = 5,\ b = 8,$ and $c = 5$.

$$x = \frac{-b \pm \sqrt{b^2 - 4ac}}{2a}$$

$$x = \frac{-8 \pm \sqrt{(8)^2 - 4(5)(5)}}{2(5)}$$

$$x = \frac{-8 \pm \sqrt{-36}}{10} = \frac{-8 \pm 6i}{10}$$

$$x = \frac{2(-4 \pm 3i)}{2(5)}$$

$$x = \frac{-4 \pm 3i}{5} = -\frac{4}{5} \pm \frac{3}{5}i$$

The solution set is $\left\{ -\dfrac{4}{5} - \dfrac{3}{5}i,\ -\dfrac{4}{5} + \dfrac{3}{5}i \right\}$.

2. Solve $2x^2 = 2x + 3$.

_____

3. Solve $(2x - 6)(x + 1) = -16$.

_____

Objective 2 Practice Exercises

For extra help, see Examples 1–3 on pages 514–516 of your text.

Use the quadratic formula to solve each equation. (All solutions for these equations are real numbers.)

1. $(z+2)^2 = 2(5z-2)$

    1. _____

2. $5k^2 + 4k - 2 = 0$

    2. _____

3. $34 - 10x = -x^2$

    3. _____

Objective 3   Use the discriminant to determine the number and type of solutions.

Video Examples

Review these examples for Objective 3:

4.   Find the discriminant. Use it to predict the number and type of solutions for each equation. Then tell whether the equation can be solved by factoring or whether the quadratic formula should be used.

a. $3x^2 + x - 2 = 0$

First identify the values of a, b, and c.
    $a = 3$, $b = 1$, and $c = -2$.
Then find the discriminant.
$$b^2 - 4ac = 1^2 - 4(3)(-2)$$
$$= 1 + 24$$
$$= 25, \text{ or } 5^2$$
Since a, b, and c are integers and the

Now Try:

4.   Find the discriminant. Use it to predict the number and type of solutions for each equation. Then tell whether the equation can be solved by factoring or whether the quadratic formula should be used.

a. $10x^2 + 21x + 9 = 0$

    _____

discriminant 25 is a perfect square, there will be two rational solutions. The equation can be solved by factoring.

b. $16x^2 + 25 = 40x$

Write in standard form: $16x^2 - 40x + 25 = 0$.
$a = 16$, $b = -40$, and $c = 25$

$$b^2 - 4ac = (-40)^2 - 4(16)(25)$$
$$= 1600 - 1600$$
$$= 0$$

Because the discriminant is 0, this quadratic equation will have one distinct rational solution. The equation can be solved by factoring.

c. $5y^2 - 5y + 2 = 0$

$a = 5$, $b = -5$, and $c = 2$

$$b^2 - 4ac = (-5)^2 - 4(5)(2)$$
$$= 25 - 40$$
$$= -15$$

Because the discriminant is negative and a, b, and c are integers, this quadratic equation will have two nonreal complex solutions. The quadratic equation should be used to solve it.

b. $25x^2 + 9 = 30x$

_____

c. $2y^2 + 4y + 8 = 0$

_____

Objective 3 Practice Exercises

For extra help, see Example 4 on pages 517–518 of your text.

Use the discriminant to determine whether the solutions for each equation are

    A. *two rational numbers*   B. *one rational number,*
    C. *two irrational numbers*   D. *two imaginary numbers.*

Do not actually solve.

4. $m^2 - 4m + 4 = 0$

4. _____

5. $z^2 + 6z + 3 = 0$

5. _____

6. $16x^2 - 12x + 9 = 0$

6. _____

Chapter 8   QUADRATIC EQUATIONS, INEQUALITIES, AND FUNCTIONS

8.3   Equations Quadratic in Form

Learning Objectives
1   Solve an equation with fractions by writing it in quadratic form.
2   Use quadratic equations to solve applied problems.
3   Solve an equation with radicals by writing it in quadratic form.
4   Solve an equation that is quadratic in form by substitution.

Key Terms

Use the vocabulary terms listed below to complete each statement in exercises 1–2.

quadratic in form   **standard form**

1.   A quadratic equation written in the form $ax^2 + bx + c = 0$, $a \neq 0$ is written in

_____.

2.   A nonquadratic equation that can be written as a quadratic equation is called

_____.

Objective 1   Solve an equation with fractions by writing it in quadratic form.

Video Examples

Review this example for Objective 1:

1.   Solve $5 + \dfrac{6}{m+1} = \dfrac{14}{m}$.

Multiply each side by the least common denominator, $m(m + 1)$. The domain must be restricted to $m \neq 0$, $m \neq -1$.

$$5 + \frac{6}{m+1} = \frac{14}{m}$$

$$m(m+1)\left(5 + \frac{6}{m+1}\right) = m(m+1)\frac{14}{m}$$

$$m(m+1)(5) + m(m+1)\frac{6}{m+1} = m(m+1)\frac{14}{m}$$

$$5m^2 + 5m + 6m = 14m + 14$$

$$5m^2 + 11m = 14m + 14$$

$$5m^2 - 3m - 14 = 0$$

$$(5m + 7)(m - 2) = 0$$

$$5m + 7 = 0 \quad \text{or} \quad m - 2 = 0$$

$$m = -\frac{7}{5} \quad \text{or} \qquad m = 2$$

The solution set is $\left\{-\dfrac{7}{5},\ 2\right\}$.

Now Try:

1.   Solve $4 - \dfrac{8}{x-1} = -\dfrac{35}{x}$.

_____

Objective 1 Practice Exercises

For extra help, see Example 1 on page 520 of your text.

Solve each equation. Check your solutions.

1. $\dfrac{5}{x} + \dfrac{1}{2x+7} = -\dfrac{2}{3}$

                                                            1. _____

2. $\dfrac{2m}{m-5} + \dfrac{7}{m+1} = 0$

                                                            2. _____

3. $1 + \dfrac{49}{2x} = \dfrac{15}{x+1}$

                                                            3. _____

Objective 2   Use quadratic equations to solve applied problems.

Video Examples

Review this example for Objective 2:

2. Amy rows her boat 6 miles upstream and then returns in $2\frac{6}{7}$ hours. The speed of the current is 2 miles per hour. How fast can she row?

Step 1 Read the problem carefully.

Step 2 Assign a variable. Let x = the rate that Amy rows in still water. The current slows Amy when she is going upstream, so Amy's rate going upstream is her rate in still water less the rate of

Now Try:

2. Mike can row 3 miles per hour in still water. It takes him 3 hours and 36 minutes to row 3 miles upstream and return. Find the speed of the current.

                          _____

the current, or $x - 2$. Similarly, the current makes Amy row faster when she is going downstream, so her downstream rate is $x + 2$.

Complete a table. Recall that $d = rt$.

	d	r	t
Upstream	6	$x - 2$	$\dfrac{6}{x-2}$
Downstream	6	$x + 2$	$\dfrac{6}{x+2}$

Step 3 Write an equation. The time upstream plus the time downstream equals the total time, $2\dfrac{6}{7}$ hours, or $\dfrac{20}{7}$ hours.

$$\frac{6}{x-2} + \frac{6}{x+2} = \frac{20}{7}$$

Step 4 Solve the equation. The LCD is $7(x-2)(x+2)$.

$$7(x-2)(x+2)\left(\frac{6}{x-2} + \frac{6}{x+2}\right)$$
$$= 7(x-2)(x+2)\left(\frac{20}{7}\right)$$

$$7(x+2)6 + 7(x-2)6 = (x-2)(x+2)20$$

$$42(x+2) + 42(x-2) = 20(x^2 - 4)$$

$$42x + 84 + 42x - 84 = 20x^2 - 80$$

$$20x^2 - 84x - 80 = 0$$

$$4(5x^2 - 21x - 20) = 0$$

$$4(5x+4)(x-5) = 0$$

$$5x + 4 = 0 \quad \text{or} \quad x - 5 = 0$$

$$x = -\frac{4}{5} \quad \text{or} \quad x = 5$$

Step 5 State the answer. The rate cannot be $-\dfrac{4}{5}$ mph, so the answer is Amy rows at 5 mph.

Step 6 Check that this value satisfies the original equation.

Name: _____ Date: _____

Instructor: _____ Section: _____

Objective 2 Practice Exercises

For extra help, see Examples 2–3 on pages 520–523 of your text.

Solve each problem. Round answers to the nearest tenth, if necessary.

4. Two pipes together can fill a large tank in 10 hours. One of the pipes, used alone, takes 15 hours longer than the other to fill the tank. How long would each pipe used alone take to fill the tank?

4. pipe 1 _____

   pipe 2 _____

5. A jet plane traveling at a constant speed goes 1200 miles with the wind, then turns around and travels for 1000 miles against the wind. If the speed of the wind is 50 miles per hour and the total flight takes 4 hours, find the speed of the plane.

5. _____

6. A man rode a bicycle for 12 miles and then hiked an additional 8 miles. The total time for the trip was 5 hours. If his rate when he was riding the bicycle was 10 miles per hour faster than his rate walking, what was each rate?

6. bike _____

   hike _____

Objective 3   Solve an equation with radicals by writing it in quadratic form.

Video Examples

Review these examples for Objective 3:

4.   Solve each equation.

a. $y = \sqrt{y+42}$

Start by squaring each side.
$$y = \sqrt{y+42}$$
$$y^2 = \left(\sqrt{y+42}\right)^2$$
$$y^2 = y+42$$
$$y^2 - y - 42 = 0$$
$$(y+6)(y-7) = 0$$
$$y+6 = 0 \quad \text{or} \quad y-7 = 0$$
$$y = -6 \quad \text{or} \quad y = 7$$

We must check all proposed solutions in the original equation because squaring each side of an equation can introduce extraneous solutions.
Check

$$y = \sqrt{y+42}$$
$$-6 \overset{?}{=} \sqrt{-6+42}$$
$$-6 \overset{?}{=} \sqrt{36}$$
$$-6 = 6 \quad \text{False}$$

$$y = \sqrt{y+42}$$
$$7 \overset{?}{=} \sqrt{7+42}$$
$$7 \overset{?}{=} \sqrt{49}$$
$$7 = 7 \quad \text{True}$$

The solution set is $\{7\}$.

b. $\sqrt{x} + 2 = x$

$$\sqrt{x} = x - 2$$
$$\left(\sqrt{x}\right)^2 = (x-2)^2$$
$$x = x^2 - 4x + 4$$
$$0 = x^2 - 5x + 4$$
$$0 = (x-1)(x-4)$$
$$x-1 = 0 \quad \text{or} \quad x-4 = 0$$
$$x = 1 \quad \text{or} \quad x = 4$$

Check each proposed solution in the original equation.

Now Try:

4.   Solve each equation.

a. $x = \sqrt{x+2}$

————————————

b. $\sqrt{2x} + 4 = x$

————————————

$$\sqrt{x}+2=x \qquad\qquad \sqrt{x}+2=x$$

$$\sqrt{1}+2\overset{?}{=}1 \qquad\qquad \sqrt{4}+2\overset{?}{=}4$$

$$1+2\overset{?}{=}1 \qquad\qquad 2+2\overset{?}{=}4$$

$$3=1 \quad \text{False} \qquad\qquad 4=4 \quad \text{True}$$

The solution set is $\{4\}$.

Objective 3 Practice Exercises

For extra help, see Example 4 on page 523 of your text.

Solve each equation. Check your solutions.

7. $\sqrt{7y-10}=y$ 7. _____

8. $x=\sqrt{\dfrac{x+3}{2}}$ 8. _____

9. $\sqrt{4x}+3=x$ 9. _____

Objective 4   **Solve an equation that is quadratic in form by substitution.**

Video Examples

Review these examples for Objective 4:

6. Solve the equation.

$$c^4 - 20c^2 + 64 = 0$$

Write this equation in quadratic form by substituting u for c^2.

$$u^2 - 20u + 64 = 0$$

$$(u - 4)(u - 16) = 0$$

$$u - 4 = 0 \quad \text{or} \quad u - 16 = 0$$
$$u = 4 \quad \text{or} \quad u = 16$$
$$c^2 = 4 \quad \text{or} \quad c^2 = 16$$
$$c = \pm 2 \quad \text{or} \quad c = \pm 4$$

The solution set is $\{-4, -2, \ 2, \ 4\}$.

7. Solve each equation.

   a. $(m + 5)^2 + 6(m + 5) + 8 = 0$

Step 1 Substitute u for $m + 5$.

$$(m + 5)^2 + 6(m + 5) + 8 = 0$$

$$u^2 + 6u + 8 = 0$$

Step 2 $(u + 4)(u + 2) = 0$

$$u + 4 = 0 \quad \text{or} \quad u + 2 = 0$$
$$u = -4 \quad \text{or} \quad u = -2$$

Step 3 $m + 5 = -4 \quad \text{or} \quad m + 5 = -2$

Step 4 $m = -9 \quad \text{or} \quad m = -7$

Step 5 Check that the solution set of the original equation is $\{-9, -7\}$.

   b. $x^{4/3} - 20x^{2/3} + 36 = 0$

Step 1 Substitute u for $x^{2/3}$.

$$x^{4/3} - 20x^{2/3} + 36 = 0$$

$$u^2 - 20u + 36 = 0$$

Step 2 $(u - 2)(u - 18) = 0$

$$u - 2 = 0 \quad \text{or} \quad u - 18 = 0$$
$$u = 2 \quad \text{or} \quad u = 18$$

Now Try:

6. Solve the equation.

$$x^4 - 5x^2 + 4 = 0$$

_____

7. Solve each equation.

   a. $(x - 5)^2 + 2(x - 5) - 35 = 0$

_____

   b. $x^{2/3} - 2x^{1/3} = 3$

_____

Step 3 $\qquad x^{2/3} = 2 \qquad$ or $\qquad x^{2/3} = 18$

Step 4 $\left(x^{2/3}\right)^{3/2} = (2)^{3/2}$ or $\left(x^{2/3}\right)^{3/2} = (18)^{3/2}$

$$x = 2\sqrt{2} \quad \text{or} \qquad x = 54\sqrt{2}$$

Step 5 Check that the solution set of the original equation is $\left\{2\sqrt{2},\ 54\sqrt{2}\right\}$.

Objective 4 Practice Exercises

For extra help, see Examples 5–7 on pages 524–527 of your text.

Solve each equation. Check your solutions.

10. $\quad 4t^4 = 21t^2 - 5$

10. _____

11. $\quad p^{4/3} - 12p^{2/3} + 27 = 0$

11. _____

12. $\quad \left(t^2 - 3t\right)^2 = 14\left(t^2 - 3t\right) - 40$

12. _____

Chapter 8   QUADRATIC EQUATIONS, INEQUALITIES, AND FUNCTIONS

8.4   Formulas and Further Applications

Learning Objectives

1   Solve formulas for variables involving squares and square roots.
2   Solve applied problems using the Pythagorean theorem.
3   Solve applied problems using area formulas.
4   Solve applied problems using quadratic functions as models.

Key Terms

Use the vocabulary terms listed below to complete each statement in exercises 1–2.

       quadratic function   **Pythagorean theorem**

1.   A function defined by $f(x) = ax^2 + bx + c$, for real numbers a, b, and c, with $a \neq 0$, is a _____ .

2.   The _____ states that the sum of the squares of the lengths of the legs of a right triangle equals the square of the length of the hypotenuse.

Objective 1   Solve formulas for variables involving squares and square roots.

Video Examples

Review these examples for Objective 1:

1. Solve the formula for the given variable.

$$y = \frac{1}{2}gt^2 \text{ for } t$$

The goal is to isolate t on one side.

$$y = \frac{1}{2}gt^2$$

$$2y = gt^2$$

$$\frac{2y}{g} = t^2$$

$$t = \pm\sqrt{\frac{2y}{g}}$$

$$t = \pm\frac{\sqrt{2y}}{\sqrt{g}} \cdot \frac{\sqrt{g}}{\sqrt{g}}$$

$$t = \pm\frac{\sqrt{2yg}}{g}$$

Now Try:

1. Solve the formula for the given variable.

$$F = \frac{mx}{t^2} \text{ for } t$$

_____

2. Solve $rk^2 - 3k = -s$ for k.

Write the equation in standard form and then use the quadratic formula to solve for k.

$$rk^2 - 3k = -s$$

$$rk^2 - 3k + s = 0$$

Let $a = r$, $b = -3$, and $c = s$.

$$k = \frac{-(-3) \pm \sqrt{(-3)^2 - 4(r)(s)}}{2r}$$

$$k = \frac{3 \pm \sqrt{9 - 4rs}}{2r}$$

The solutions are $\dfrac{3 + \sqrt{9 - 4rs}}{2r}$ and

$\dfrac{3 - \sqrt{9 - 4rs}}{2r}$.

2. Solve $p^2q^2 + pkq = k^2$ for q.

_____

Objective 1 Practice Exercises

For extra help, see Examples 1–2 on pages 532–533 of your text.

Solve each equation for the indicated variable. (Leave ± in your answers.)

1. $F = \dfrac{kl}{\sqrt{d}}$ for d

1. _____

2. $p = \sqrt{\dfrac{kl}{g}}$ for k

2. _____

3. $b^2a^2 + 2bca = c^2$ for a

3. _____

Objective 2   Solve applied problems using the Pythagorean theorem.

Video Examples

Review this example for Objective 2:

3. A 13-foot ladder is leaning against a building. The distance from the bottom of the ladder to the building is 2 feet more than twice the distance from the top of the ladder to the ground. How far is the bottom of the ladder from the building?

Step 1 Read the problem carefully.

Step 2 Assign a variable. Let x = the distance from the top of the ladder to the ground.
Then $2x + 2$ = the distance from the bottom of the ladder to the building. Draw a picture to represent the problem.

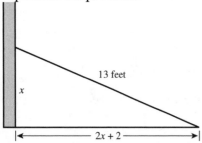

Step 3 Write an equation. Use the Pythagorean theorem.

$$a^2 + b^2 = c^2$$

$$x^2 + (2x + 2)^2 = 13^2$$

Step 4 Solve.

$$x^2 + 4x^2 + 8x + 4 = 169$$

$$5x^2 + 8x - 165 = 0$$

$$(5x + 33)(x - 5) = 0$$

$$5x + 33 = 0 \qquad \text{or} \quad x - 5 = 0$$

$$x = -\frac{33}{5} \quad \text{or} \qquad x = 5$$

Step 5 State the answer. Length cannot be negative, so discard the negative solution. The distance from the top of the ladder to the ground is 5 feet. However, we are asked to find the distance from the bottom of the ladder to the building. This distance is $2(5) + 2 = 12$ feet.

Step 6 Check. Since $5^2 + 12^2 = 13^2$, the answer is correct.

Now Try:

3. Two cars left an intersection at the same time, one heading south, the other heading east. Sometime later, the car traveling south had gone 18 miles farther than the car headed east. At that time they were 90 miles apart. How far had each car traveled?

south _____

east   _____

Objective 2 Practice Exercises

For extra help, see Example 3 on page 533 of your text.

Solve each problem.

4. A child flying a kite has let out 45 feet of string to
    the kite. The distance from the kite to the ground is 9
    feet more than the distance from the child to a point
    directly below the kite. How high up is the kite?

4. _____

5. A ladder is leaning against a building so that the top
    is 8 feet above the ground. The length of the ladder
    is 2 feet less than twice the distance of the bottom of
    the ladder from the building. Find the length of the
    ladder.

5. _____

6. Two cars left an intersection at the same time, one
    heading north, the other heading west. Later they
    were exactly 95 miles apart. The car headed west
    had gone 38 miles less than twice as far as the car
    headed north. How far had each car traveled?

6. north _____

   west _____

Objective 3   Solve applied problems using area formulas.

Video Examples

Review this example for Objective 3:

4. A fish pond is 3 feet by 4 feet. How wide a strip of concrete can be poured around the pond if there is enough concrete for 44 square feet?

Step 1 Read the problem carefully.

Step 2 Assign a variable. Let x = the width of the strip of concrete. Then $2x + 3$ = the width of the fish pond with the two strips of concrete and $2x + 4$ = the length of the fish pond with the two strips of concrete.

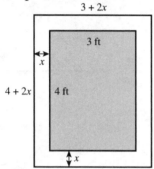

Step 3 Write an equation. The area of the strip is 44 sq ft and the area of the fish pond is $3(4) = 12$ sq ft, so the total area of the outer rectangle is $44 + 12 = 56$ sq ft.

$$(3 + 2x)(4 + 2x) = 56$$

Step 4 Solve.

$$(3 + 2x)(4 + 2x) = 56$$
$$12 + 14x + 4x^2 = 56$$
$$4x^2 + 14x - 44 = 0$$
$$2(2x^2 + 7x - 22) = 0$$
$$2(2x + 11)(x - 2) = 0$$
$$2x + 11 = 0 \quad \text{or} \quad x - 2 = 0$$
$$x = -\frac{11}{2} \quad \text{or} \quad x = 2$$

Step 5 State the answer. Since length cannot be negative, we disregard the negative solution. The concrete strip should be 2 ft wide.

Step 6 Check. If $x = 2$, then the area of the large rectangle is $(3 + 2 \cdot 2)(4 + 2 \cdot 2) = 7(8) = 56$ sq ft. The area of the fish pond is $3(4) = 12$ sq ft, so the area of the concrete strip is $56 - 12 = 44$ sq ft.

Now Try:

4. A picture 9 inches by 12 inches is to be mounted on a piece of mat board so that there is an even width of mat all around the picture. How wide will the matted border be if the area of the mounted picture is 238 square inches?

_____

Objective 3 Practice Exercises

For extra help, see Example 4 on page 534 of your text.

Solve each problem.

7.   A rug is to fit in a room so that a border of even width is left on all four sides. If the room is 16 feet by 20 feet and the area of the rug is 165 square feet, how wide to the nearest tenth of a foot will the border be?

7. _____

8.   A rectangular garden has an area of 12 feet by 5 feet. A gravel path of equal width is to be built around the garden. How wide can the path be if there is enough gravel for 138 square feet?

8. _____

9.   A doghouse 2 feet by 4 feet is to be built with a cement path around it of equal width on all sides. The area available for the doghouse and path is 120 square feet. How wide will the path be?

9. _____

Objective 4   Solve applied problems using quadratic functions as models.

Video Examples

Review this example for Objective 4:

5.   A certain projectile is located at a distance of
      $d(t) = 3t^2 - 6t + 1$ feet from its starting point
      after t seconds. How many seconds will it take
      the projectile to travel 10 feet?

      Let $d = 10$ in the formula and solve for t.

$$d = 3t^2 - 6t + 1$$
$$10 = 3t^2 - 6t + 1$$
$$9 = 3t^2 - 6t$$
$$3 = t^2 - 2t$$
$$3 + 1 = t^2 - 2t + 1$$
$$4 = (t-1)^2$$
$$\sqrt{4} = t - 1 \quad \text{or} \quad -\sqrt{4} = t - 1$$
$$2 = t - 1 \quad \text{or} \quad -2 = t - 1$$
$$3 = t \quad \text{or} \quad -1 = t$$

      Since t represents time, we reject the negative
      solution. It will take 3 seconds for the projectile
      to travel 10 feet.

Now Try:

5.   A baseball is thrown upward
      from a building 20 m high with
      a velocity of 15 m/sec. Its
      distance from the ground after t
      seconds is modeled by the
      function
      $$f(t) = -4.9t^2 + 15t + 20.$$
      When will the ball hit the
      ground? Round your answer to
      the nearest tenth.

      _____

For extra help, see Examples 5–6 on pages 535–536 of your text.

Solve each problem. Round answers to the nearest tenth.

10.   A population of microorganisms grows according to   10.   _____
       the function $p(x) = 100 + 0.2x + 0.5x^2$, where x is
       given in hours. How many hours does it take to
       reach a population of 250 microorganisms?

11. An object is thrown downward from a tower 280 feet   **11.** _____
high. The distance the object has fallen at time t in
seconds is given by $s(t) = 16t^2 + 68t$. How long will
it take the object to fall 100 feet?

12. A widget manufacturer estimates that her monthly   **12.** _____
revenue can be modeled by the function
$R(x) = -0.006x^2 + 32x - 10,000$. What is the
minimum number of items that must be sold for the
revenue to equal $30,000?

Name: Date:
Instructor: Section:

Chapter 8 QUADRATIC EQUATIONS, INEQUALITIES, AND FUNCTIONS

8.5   Graphs of Quadratic Functions

Learning Objectives
1   Graph a quadratic function.
2   Graph parabolas with horizontal and vertical shifts.
3   Use the coefficient of x^2 to predict the shape and direction in which a parabola opens.
4   Find a quadratic function to model data.

Key Terms

Use the vocabulary terms listed below to complete each statement in exercises 1−4.

     parabola   vertex   axis   quadratic function

1.   The vertical (or horizontal) line through the vertex of a vertical (or horizontal) parabola is its _____.

2.   The point on a parabola that has the least y-value (if the parabola opens up) or the greatest y-value (if the parabola opens down) is called the _____ of the parabola.

3.   A function defined by $f(x) = ax^2 + bx + c$, for real numbers a, b, and c, with $a \neq 0$, is a _____.

4.   The graph of a quadratic function is a _____.

Objective 1   Graph a quadratic function.

For extra help, see page 541 of your text.

Objective 2   Graph parabolas with horizontal and vertical shifts.

Video Examples

Review these examples for Objective 2:

1.   Graph $g(x) = x^2 + 2$. Give the vertex, axis, domain, and range.

The graph of $g(x)$ has the same shape as that of $f(x) = x^2$ but shifted 2 units up with vertex (0, 2). Every function value is 2 more than the corresponding function value of $f(x) = x^2$.

Now Try:

1.   Graph $f(x) = x^2 - 1$. Give the vertex, axis, domain, and range.

Vertex   _____

Axis   _____

Domain   _____

Range   _____

Copyright © 2016 Pearson Education, Inc.

x	$f(x) = x^2$	$g(x) = x^2 + 2$
-2	4	6
-1	1	3
0	0	2
1	1	3
2	4	6

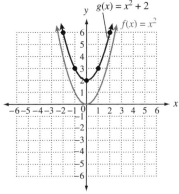

The vertex is (0, 2). The axis is $x = 0$. The domain is $(-\infty, \infty)$. The range is $[2, \infty)$.

2. Graph $g(x) = (x-1)^2$. Give the vertex, axis, domain, and range.

The graph of $g(x)$ has the same shape as that of $f(x) = x^2$ but shifted 1 unit right with vertex (1, 0).

x	$f(x) = x^2$	$g(x) = (x-1)^2$
-2	4	9
-1	1	4
0	0	1
1	1	0
2	4	1
3	9	4

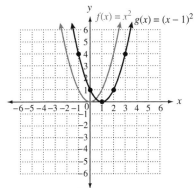

The vertex is (1, 0). The axis is $x = 1$. The domain is $(-\infty, \infty)$. The range is $[0, \infty)$.

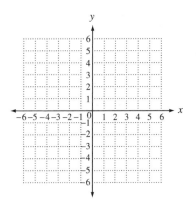

2. Graph $f(x) = (x-3)^2$. Give the vertex, axis, domain, and range.

Vertex   _____

Axis   _____

Domain   _____

Range   _____

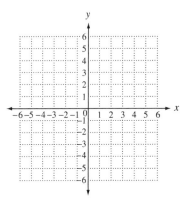

3. Graph $g(x) = (x-1)^2 - 2$. Give the vertex, axis, domain, and range.

The graph of $g(x)$ has the same shape as that of $f(x) = x^2$ but shifted 1 unit right (since $x - 1 = 0$ if $x = 1$) and 2 units down (because of the -2).

x	$g(x) = (x-1)^2 - 2$
-2	7
-1	2
0	-1
1	-2
2	-1
3	2
4	7

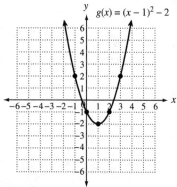

The vertex is $(1, -2)$. The axis is $x = 1$. The domain is $(-\infty, \infty)$. The range is $[-2, \infty)$.

3. Graph $f(x) = (x+2)^2 - 1$. Give the vertex, axis, domain, and range.

Vertex _____

Axis _____

Domain _____

Range _____

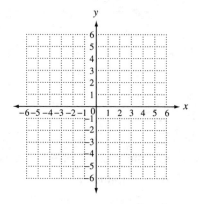

Objective 2 Practice Exercises

For extra help, see Examples 1–3 on pages 542–543 of your text.

Sketch the graph of each parabola. Give the vertex, axis, domain, and range.

1. $f(x) = x^2 - 4$

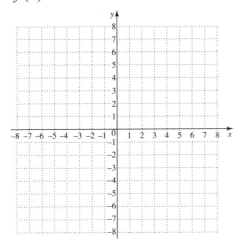

1. vertex _____

axis _____

domain _____

range _____

2. $f(x) = (x-3)^2$

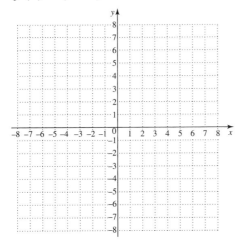

2. vertex _____

axis _____

domain _____

range _____

3. $f(x) = (x-3)^2 - 1$

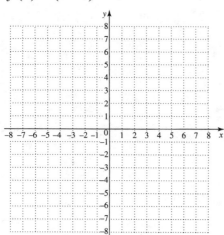

3. vertex _____

axis _____

domain _____

range _____

Objective 3 Use the coefficient of x^2 to predict the shape and direction in which a parabola opens.

Video Examples

Review these examples for Objective 3:

4. Graph $g(x) = -2x^2$. Give the vertex, axis, domain, and range.

The graph of $g(x)$ has the same shape as that of $f(x) = x^2$ but is narrower and opens downward.

x	$g(x) = -2x^2$
-2	-8
-1	-2
0	0
1	-2
2	-8

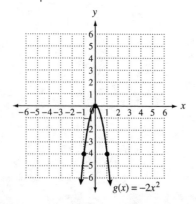

The vertex is $(0, 0)$. The axis is $x = 0$. The domain is $(-\infty, \infty)$. The range is $(-\infty, 0]$.

Now Try:

4. Graph $g(x) = -\frac{1}{4}x^2$. Give the vertex, axis, domain, and range.

Vertex _____

Axis _____

Domain _____

Range _____

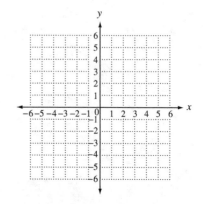

Name: Date:

Instructor: Section:

5. Graph $g(x) = -\frac{1}{2}(x+1)^2 - 2$. Give the vertex, axis, domain, and range.

The parabola opens down because $a < 0$ and is wider than the graph of $f(x) = x^2$. The parabola has vertex $(-1,-2)$.

x	$g(x) = -\frac{1}{2}(x+1)^2 - 2$
-3	-4
-2	-2.5
-1	-2
0	-2.5
1	-4

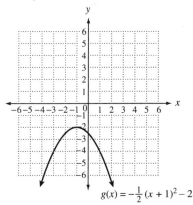

$$g(x) = -\tfrac{1}{2}(x+1)^2 - 2$$

The vertex is $(-1,-2)$. The axis is $x = -1$. The domain is $(-\infty, \infty)$. The range is $(-\infty, -2]$.

5. Graph $f(x) = 3(x-1)^2 + 1$. Give the vertex, axis, domain, and range.

Vertex   _____

Axis   _____

Domain   _____

Range   _____

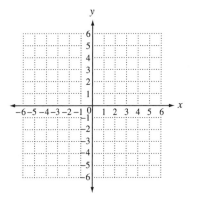

Objective 3 Practice Exercises

For extra help, see Examples 4–5 on page 544 of your text.

For each quadratic function, tell whether the graph opens up or down and whether the graph is wider, narrower, or the same shape as the graph of $f(x) = x^2$. Then give the vertex, domain, and range.

4. $f(x) = -\frac{4}{3}x^2 - 1$

4.   _____

    vertex _____

    domain _____

    range _____

5. $f(x) = -2(x+1)^2$

5. _____

vertex _____

domain _____

range _____

6. $f(x) = \frac{5}{4}(x-1)^2 + 7$

6. _____

vertex _____

domain _____

range _____

Objective 4   Find a quadratic function to model data.

Video Examples

Review this example for Objective 4:

6. The number of ice cream cones sold by an ice cream parlor from 2003–2009 is shown in the following table.

Year	Years since 2003, x	Number of cones sold
2003	0	1775
2004	1	4194
2005	2	5063
2006	3	5161
2007	4	4663
2008	5	4639
2009	6	3710

Use the ordered pairs (x, number of cones sold) to make a scatter diagram of the data.
Determine a quadratic function that models these data by using a system of equations. Use the ordered pairs (0, 1775), (3, 5161), and (6, 3710). Round the values of a, b, and c in your model to the nearest tenth, as necessary.

Now Try:

6. The table lists the average price of a Major League Baseball ticket.

Year	Years since 1990, x	Price
1991	1	$9.14
1994	4	$10.60
1997	7	$12.49
2000	10	$16.81
2004	14	$19.82
2010	20	$26.74

Use the ordered pairs (x, price) to make a scatter diagram of the data. Determine a quadratic function that models these data by using a system of equations. Use the ordered pairs (1, 9.14), (10, 16.81), and (20, 26.74). Round the values of a, b, and c in your model to the nearest hundredth, as necessary.

_____

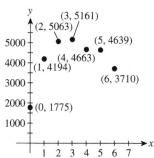

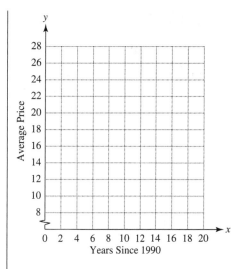

It appears that the parabola opens down, so the coefficient a is negative.

Using the chosen ordered pairs, we substitute the x- and y-values into the quadratic form

$y = ax^2 + bx + c$ to obtain the three equations.

$a(0)^2 + b(0) + c = 1775$ (1)

$a(3)^2 + b(3) + c = 5161$ (2)

$a(6)^2 + b(6) + c = 3710$ (3)

Equation (1) simplifies to $c = 1775$, so substitute 1775 for c in equation (2) and (3).

$9a + 3b + 1775 = 5161$ (2)

$36a + 6b + 1775 = 3710$ (3)

Subtract 1775 from each side of both equations.

$9a + 3b = 3386$ (2)

$36a + 6b = 1935$ (3)

Solve by elimination. Multiply equation (2) by -2 and add to equation (3).

$-18a - 6b = -6772$ Multiply (2) by -2.

$\underline{36a + 6b = 1935}$ (3)

$18a = -4837$

$a \approx -268.7$ Round to the tenth.

Substitute this value for a into equation (2) and solve for b.

$9a + 3b = 3386$ (2)

$9(-268.7) + 3b = 3386$

$-2418.3 + 3b = 3386$

$3b = 5804.3$

$b \approx 1934.8$

Therefore, the model

is $y = -268.7x^2 + 1934.8x + 1775$.

Objective 4 Practice Exercises

For extra help, see Example 6 on pages 545–546 of your text.

Tell whether a linear or quadratic function would be a more appropriate model for each set of graphed data. If linear, tell whether the slope should be positive or negative. If quadratic, tell whether the coefficient a of x^2 should be positive or negative.

7.

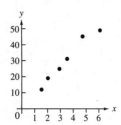

7. _____

8.

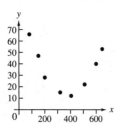

8. _____

Solve the problem.

9.   The number of publicly traded companies filing for bankruptcy for selected years between 1990 and 2000 are shown in the table, with 0 representing 1990, 2 representing 1992, etc.

Year	Number of Bankruptcies
0	115
2	91
4	70
6	84
8	120
10	176

9.

Use the ordered pairs to make a scatter diagram of the data.

Use the ordered pairs (0, 115), (4, 70), and (8, 120) to find a function that models the data. Round the values of *a*, *b*, and *c* to three decimal places, if necessary.

Source: Lial, Margaret L., John Hornsby, Terry McGinnis, *Intermediate Algebra* Eighth Edition. Boston: Pearson Education, 2006.

Chapter 8   QUADRATIC EQUATIONS, INEQUALITIES, AND FUNCTIONS

8.6   More about Parabolas and Their Applications

Learning Objectives

1	Find the vertex of a vertical parabola.
2	Graph a quadratic function.
3	Use the discriminant to find the number of x-intercepts of a parabola with a vertical axis.
4	Use quadratic functions to solve problems involving maximum or minimum value.
5	Graph parabolas with horizontal axes.

Key Terms

Use the vocabulary terms listed below to complete each statement in exercises 1−2.

discriminant   **vertex**

1. The_____ of a quadratic function is found by using the formula $b^2 - 4ac$.

2. The maximum or minimum value of a quadratic function occurs at the _____ of its graph.

Objective 1   Find the vertex of a vertical parabola.

Video Examples

Review these examples for Objective 1:

1. Find the vertex of the graph of
$f(x) = x^2 + 6x + 10$.

We can express $x^2 + 6x + 10$ in the form $(x-h)^2 + k$ by completing the square on $x^2 + 6x$. Because we want to keep $f(x)$ alone on one side of the equation, we add and subtract the appropriate number on just one side.

$$f(x) = x^2 + 6x + 10 \qquad \left[\frac{1}{2}(6)\right]^2 = 9$$

$$f(x) = (x^2 + 6x + 9 - 9) + 10$$

$$f(x) = (x^2 + 6x + 9) + 10 - 9$$

$$f(x) = (x+3)^2 + 1$$

The vertex of the parabola is (−3, 1).

Now Try:

1. Find the vertex of the graph of
$f(x) = x^2 - 6x + 4$.

_____

2. Find the vertex of the graph of
$f(x) = 3x^2 + 6x + 10.$

Because the x^2-term has a coefficient other than 1, we factor that coefficient out of the first two terms before completing the square.

$$f(x) = 3x^2 + 6x + 10$$

$$f(x) = 3(x^2 + 2x) + 10 \qquad \left[\frac{1}{2}(2)\right]^2 = 1$$

$$f(x) = 3(x^2 + 2x + 1 - 1) + 10$$

$$f(x) = 3(x^2 + 2x + 1) + 3(-1) + 10$$

$$f(x) = 3(x+1)^2 + 7$$

The vertex is (−1, 7).

2. Find the vertex of the graph of
$f(x) = -2x^2 + 4x - 1.$

_____

3. Use the vertex formula to find the vertex of the graph of $f(x) = -4x^2 + 5x + 3.$

The x-coordinate of the vertex of the parabola is given by $\dfrac{-b}{2a}$.

$$\frac{-b}{2a} = \frac{-5}{2(-4)} = \frac{5}{8}$$

The y-coordinate is $f\left(\dfrac{-b}{2a}\right) = f\left(\dfrac{5}{8}\right)$.

$$f\left(\frac{5}{8}\right) = -4\left(\frac{5}{8}\right)^2 + 5\left(\frac{5}{8}\right) + 3 = \frac{73}{16}$$

The vertex is $\left(\dfrac{5}{8}, \dfrac{73}{16}\right)$.

3. Use the vertex formula to find the vertex of the graph of
$f(x) = 2x^2 - 6x + 5.$

_____

Objective 1 Practice Exercises

For extra help, see Examples 1–3 on pages 550–552 of your text.

Find the vertex of each parabola.

1. $f(x) = x^2 - 2x + 4$

1. _____

2. $f(x) = 5x^2 - 4x + 1$

2. _____

3. $f(x) = -\dfrac{1}{4}x^2 - 3x - 9$

3. _____

Name: Date:

Instructor: Section:

Objective 2   Graph a quadratic function.

Video Examples

Review this example for Objective 2:

4. Graph the quadratic function defined by $f(x) = x^2 - 3x + 2$. Give the vertex, axis, domain, and range.

    Step 1 From the equation, $a = 1$, so the graph opens up.

    Step 2 The x-coordinate of the vertex is $\frac{3}{2}$. The y-coordinate of the vertex is

    $f\left(\frac{3}{2}\right) = \left(\frac{3}{2}\right)^2 - 3\left(\frac{3}{2}\right) + 2 = -\frac{1}{4}$. The vertex is $\left(\frac{3}{2}, -\frac{1}{4}\right)$.

    Step 3 Find any intercepts. Since the vertex is in quadrant IV and the graph opens up, there will be two x-intercepts. Let $f(x) = 0$ and solve.

    $$x^2 - 3x + 2 = 0$$
    $$(x-1)(x-2) = 0$$
    $$x - 1 = 0 \quad \text{or} \quad x - 2 = 0$$
    $$x = 1 \quad \text{or} \quad x = 2$$

    The x-intercepts are $(1, 0)$ and $(2, 0)$. Find the y-intercept by evaluating $f(0)$.

    $$f(0) = 0^2 - 3(0) + 2$$

    The y-intercept is $(0, 2)$.

    Step 4 Plot the points found so far and additional points as needed using symmetry about the axis, $x = \frac{3}{2}$.

    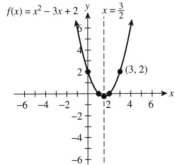

    The domain is $(-\infty, \infty)$. The range is $\left[-\frac{1}{4}, \infty\right)$.

Now Try:

4. Graph the quadratic function defined by $f(x) = x^2 + 4x + 5$. Give the vertex, axis, domain, and range.

    Vertex   _____

    Axis   _____

    Domain   _____

    Range   _____

    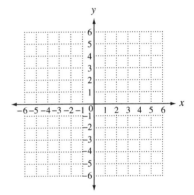

Name: Date:

Instructor: Section:

Objective 2 Practice Exercises

For extra help, see Example 4 on pages 552–553 of your text.

Sketch the graph of each parabola. Give the vertex, axis, domain, and range.

4. $f(x) = -x^2 + 8x - 10$

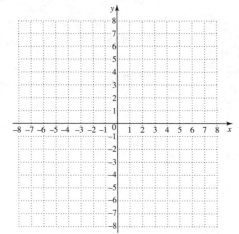

4. vertex _____

      axis _____

      domain _____

      range _____

5. $f(x) = 3x^2 + 6x + 2$

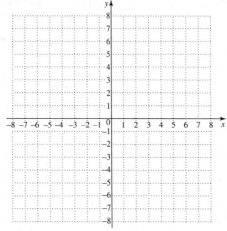

5. vertex _____

      axis _____

      domain _____

      range _____

6. $f(x) = -2x^2 + 4x + 1$

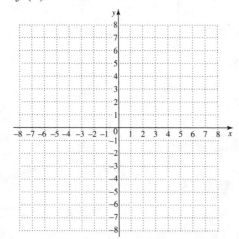

6. vertex _____

      axis _____

      domain _____

      range _____

Objective 3　Use the discriminant to find the number of *x*-intercepts of a parabola with a vertical axis.

Video Examples

Review this example for Objective 3:

5. Use the discriminant to determine the number of *x*-intercepts of the graph of the quadratic function.

$$f(x) = 4x^2 + 12x + 9$$

$$b^2 - 4ac = 12^2 - 4(4)(9) = 0$$

Since the discriminant is zero, the graph has only one *x*-intercept, its vertex.

Now Try:

5. Use the discriminant to determine the number of *x*-intercepts of the graph of the quadratic function.

$$f(x) = 9x^2 - 24x + 16$$

_____

Objective 3 Practice Exercises

For extra help, see Example 5 on pages 553–554 of your text.

Use the discriminant to determine the number of x-intercepts of the graph of each function.

7.　$f(x) = 2x^2 - 3x + 2$

7. _____

8.　$f(x) = -3x^2 - x + 5$

8. _____

9.　$f(x) = 3x^2 - 6x + 3$

9. _____

Objective 4　Use quadratic functions to solve problems involving maximum or minimum value.

Video Examples

Review this example for Objective 4:

6. A farmer has 1000 yards of fencing to enclose a rectangular field. What is the largest area that the farmer can enclose? What are the dimensions of the field when the area is maximized?

If the length of the field is represented by *l* and the width of the field is represented by *w*, the perimeter is given by $2l + 2w = 1000$ or $l + w = 500$ or $w = 500 - l$.

Now Try:

6. A farmer has 1000 yards of fencing to enclose a rectangular field next to a building. What is the largest area that the farmer can enclose? What are the dimensions of the field when the area is maximized?

_____

The area is given by $A = lw$ or

$$A(l) = l(500 - l) = 500l - l^2 = -l^2 + 500l.$$

This is a quadratic equation, so its maximum occurs at the vertex of its graph. The

x-coordinate is given by $\dfrac{-b}{2a} = \dfrac{-500}{2(-1)} = 250.$

The y-coordinate is

$$A(250) = -250^2 + 500(250) = 62,500.$$

If $l = 250$, then $w = 500 - 250 = 250$.

Therefore, the maximum area is 62,500 sq yd when the length of the field is 250 yd and the width is 250 yd.

Objective 4 Practice Exercises

For extra help, see Examples 6–7 on pages 554–555 of your text.

Solve each problem.

10. Jean sells ceramic pots. She has weekly costs of $C(x) = x^2 - 100x + 2700,$ where x is the number of pots she sells each week. How many pots should she sell to minimize her costs? What is the minimum cost?

10. units _____

    cost _____

11. The length and width of a rectangle have a sum of 48. What width will produce the maximum area?

11. _____

12. A projectile is fired upward so that its distance (in feet) above the ground t seconds after firing is given by $s(t) = -16t^2 + 80t + 156$. Find the maximum height it reaches and the number of seconds it takes to reach that height.

12. height _____

time _____

Objective 5   Graph parabolas with horizontal axes.

Video Examples

Review this example for Objective 5:

9. Graph $x = -y^2 + 6y - 9$. Give the vertex, axis, domain, and range.

We must complete the square in order to write the equation in $x = (y - k)^2 + h$ form.

$$x = -(y^2 - 6y) - 9$$

$$x = -(y^2 - 6y + 9 - 9) - 9$$

$$x = -(y^2 - 6y + 9) - 1(-9) - 9$$

$$x = -(y - 3)^2$$

The vertex is (0, 3). The axis is $y = 3$.

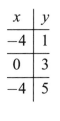

x	y
-4	1
0	3
-4	5

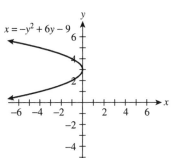

domain: $(-\infty, 0]$

range: $(-\infty, \infty)$

Now Try:

9. Graph $x = -y^2 + 4y - 4$. Give the vertex, axis, domain, and range.

Vertex _____

Axis _____

Domain _____

Range _____

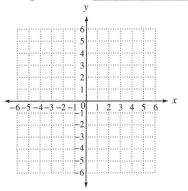

Name: Date:
Instructor: Section:

Objective 5 Practice Exercises

For extra help, see Examples 8–9 on page 556 of your text.

Sketch the graph of each parabola. Give the vertex, axis, domain, and range.

13. $x = -y^2 + 2$

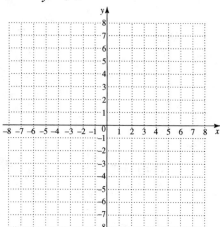

13. vertex _____

     axis _____

     domain _____

     range _____

14. $x = y^2 - 3$

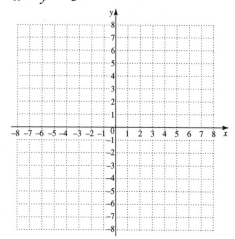

14. vertex _____

     axis _____

     domain _____

     range _____

15. $x = -y^2 - 6y - 10$

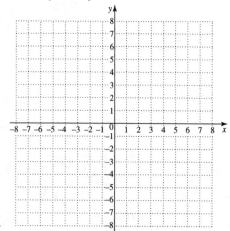

15. vertex _____

     axis _____

     domain _____

     range _____

Chapter 8   QUADRATIC EQUATIONS, INEQUALITIES, AND FUNCTIONS

8.7   Polynomial and Rational Inequalities

Learning Objectives

1   Solve quadratic inequalities.
2   Solve polynomial inequalities of degree 3 or greater.
3   Solve rational inequalities.

Key Terms

Use the vocabulary terms listed below to complete each statement in exercises 1–2.

    quadratic inequality   **rational inequality**

1.   An inequality that involves a rational expression is a _____.

2.   An inequality that can be written in the form $ax^2 + bx + c < 0$ or $ax^2 + bx + c > 0$, where a, b, and c are real numbers with $a \neq 0$ is called a
_____.

Objective 1   Solve quadratic inequalities.

Video Examples

Review these examples for Objective 1:

1.   Solve each inequality.

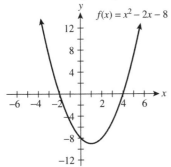

  a. $x^2 - 2x - 8 > 0$

From the graph, we see that the y-values are greater than 0 when the x-values are less than –2 or greater than 4. Therefore, the solution set of $x^2 - 2x - 8 > 0$ is $(-\infty, -2) \cup (4, \infty)$.

  b. $x^2 - 2x - 8 < 0$

From the graph, we see that the y-values are less than 0 when the x-values are greater than –2 and less than 4. Therefore, the solution set of $x^2 - 2x - 8 < 0$ is $(-2, 4)$.

Now Try:

1.   Solve each inequality.

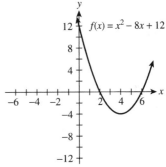

  a. $x^2 - 8x + 12 > 0$

_____

  b. $x^2 - 8x + 12 < 0$

_____

2. Solve and graph the solution set of
$x^2 + 5x + 4 \geq 0$.

Solve the quadratic equation by factoring.
$(x+1)(x+4) = 0$
$x+1 = 0$ or $x+4 = 0$
   $x = -1$ or $x = -4$

The numbers –4 and –1 divide a number line into intervals A, B, and C, as shown below.

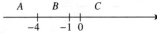

Since the numbers –4 and –1 are the only numbers that make the quadratic expression

$x^2 + 5x + 4$ equal to 0, all other numbers make the expression either positive or negative. If one number in an interval satisfies the inequality, then all the numbers in that interval will satisfy the inequality.

Choose any number in interval A as a test number; we will choose –5.

$$x^2 + 5x + 4 \geq 0$$
$$\overset{?}{(-5)^2 + 5(-5) + 4 \geq 0}$$
$$4 \geq 0 \quad \text{True}$$

Because –5 satisfies the inequality, all numbers from interval A are solutions.

Now try –2 from interval B.

$$x^2 + 5x + 4 \geq 0$$
$$\overset{?}{(-2)^2 + 5(-2) + 4 \geq 0}$$
$$-2 \geq 0 \quad \text{False}$$

The numbers in interval B are not solutions.

Finally, try 0 from interval C.

$$x^2 + 5x + 4 \geq 0$$
$$\overset{?}{0^2 + 5(0) + 4 \geq 0}$$
$$4 \geq 0 \quad \text{True}$$

Because 0 satisfies the inequality, all numbers from interval C are solutions.

Because the inequality is greater than or equal to zero, we include the endpoints of the intervals in the solution set. Thus, the solution set is

$(-\infty, -4] \cup [-1, \infty)$.

2. Solve and graph the solution set
of $x^2 - x - 2 < 0$.

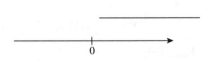

Name: Date:
Instructor: Section:

4. Solve each inequality.

  a. $(2k+5)^2 \geq -1$

Because $(2k+5)^2$ is never negative, it is always greater than -1. The solution set is $(-\infty, \infty)$.

  b. $(2k+5)^2 \leq -1$

Because $(2k+5)^2$ is never negative, there is no solution. The solution set is $\varnothing$.

4. Solve each inequality.

  a. $(4m+1)^2 \geq -3$

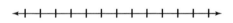

  b. $(4m+1)^2 \leq -3$

Objective 1 Practice Exercises

For extra help, see Examples 1–4 on pages 562–565 of your text.

Solve each inequality, and graph the solution set.

1. $a^2 - a - 2 \leq 0$

1. _____

                                           ←++++++++++++++++→

2. $8k^2 + 10k > 3$

2. _____

                                           ←++++++++++++++++→

3. $(3x-2)^2 < -1$

3. _____

                                           ←++++++++++++++++→

Name: Date:

Instructor: Section:

Objective 2   **Solve polynomial inequalities of degree 3 or greater.**

Video Examples

Review this example for Objective 2:

5.   Solve and graph the solution set of
$(x+1)(x-2)(x+4) \leq 0$.

Set the factored polynomial equal to 0, then use the zero-factor property.

$x+1=0$ or $x-2=0$ or $x+4=0$

   $x=-1$ or $x=2$ or $x=-4$

Locate –4, –1, and 2 on a number line to determine the intervals A, B, C, and D.

```
   A     B   C      D
 --+-----+---+--+------->
  -4    -1 0  2
```

Substitute a test number from each interval in the original inequality to determine which intervals satisfy the inequality.

Interval	Test Number	Test of inequality	True or False?
A	-5	$-28 \leq 0$	T
B	-2	$8 \leq 0$	F
C	0	$-8 \leq 0$	T
D	5	$162 \leq 0$	F

The numbers in intervals A and C are in the solution set. The three endpoints are included in the solution set since the inequality symbol, $\leq$, includes equality. Thus, the solution set is $(-\infty, -4] \cup [-1, \ 2]$.

```
 <----]     [--+--]     --->
     -4    -1 0  2
```

Now Try:

5.   Solve and graph the solution set of $(2x-1)(2x+3)(3x+1) \leq 0$.

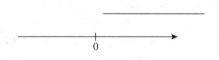

Objective 2 Practice Exercises

For extra help, see Example 5 on pages 565–566 of your text.

Solve each inequality, and graph the solution set.

4. $(y+2)(y-1)(y-2) < 0$

4. _____

```
 <-+--+--+--+--+--+--+--+--+--+--+--+->
```

5. $(k+5)(k-1)(k+3) \le 0$

5. _____

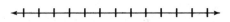

6. $(x-1)(x-3)(x+2) \ge 0$

6. _____

Objective 3   **Solve rational inequalities.**

Video Examples

Review these examples for Objective 3:

6. Solve and graph the solution set of $\dfrac{7}{x-1} < 1$.

Write the inequality so that 0 is on one side.

$$\frac{7}{x-1} - 1 < 0$$

$$\frac{7}{x-1} - \frac{x-1}{x-1} < 0 \quad \text{The LCD is } x-1.$$

$$\frac{7-x+1}{x-1} < 0$$

$$\frac{8-x}{x-1} < 0$$

The sign of $\dfrac{8-x}{x-1}$ will change from positive to negative or negative to positive only at those numbers that make the numerator or denominator 0. These two numbers, 1 and 8, divide a number line into three intervals.

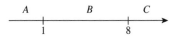

Test a number in each interval using the original inequality.

Interval	Test Number	Test of inequality	True or False?
A	0	$-7 < 1$	T
B	2	$7 < 1$	F
C	10	$\dfrac{7}{9} < 1$	T

The solution set is $(-\infty, 1) \cup (8, \infty)$. This interval does not include 1 because it would

Now Try:

6. Solve and graph the solution set of $\dfrac{y}{y+1} > 3$.

_____

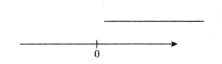

make the denominator of the original inequality 0. The number 8 is not included because the inequality symbol, <, does not include equality.

7. Solve and graph the solution set of $\dfrac{x+1}{x-5} \ge 3$.

Write the inequality so that 0 is on one side.

$$\frac{x+1}{x-5} - 3 \ge 0$$

$$\frac{x+1}{x-5} - \frac{3(x-5)}{x-5} \ge 0$$

$$\frac{x+1}{x-5} - \frac{3x-15}{x-5} \ge 0$$

$$\frac{x+1-3x+15}{x-5} \ge 0$$

$$\frac{-2x+16}{x-5} \ge 0$$

The sign of $\dfrac{-2x+16}{x-5}$ will change from positive to negative or negative to positive only at those numbers that make the numerator or denominator 0. These two numbers, 5 and 8, divide a number line into three intervals.

Test a number in each interval using original inequality.

Interval	Test Number	Test of inequality	True or False?
A	0	$-\dfrac{1}{5} \ge 3$	F
B	6	$7 \ge 3$	T
C	10	$\dfrac{11}{5} \ge 3$	F

The solution set is $(5, \, 8]$. This interval does not include 5 because it would make the denominator of the original inequality 0. The number 8 is included because the inequality symbol, $\ge$, does includes equality.

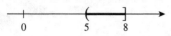

7. Solve and graph the solution set of $\dfrac{z+2}{z-3} \le 2$.

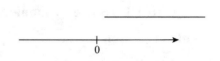

Name: Date:

Instructor: Section:

For extra help, see Examples 6–7 on pages 566–568 of your text.

Solve each inequality, and graph the solution set.

7. $\dfrac{7}{x-1} \leq 1$

                                                 7. _____

8. $\dfrac{2p-1}{3p+1} \leq 1$

                                                 8. _____

9. $\dfrac{5}{x-3} \leq -1$

                                                 9. _____

Chapter 9   INVERSE, EXPONENTIAL, AND LOGARITHMIC FUNCTIONS

9.1   Inverse Functions

Learning Objectives
1   Decide whether a function is one-to-one and, if it is, find its inverse.
2   Use the horizontal line test to determine whether a function is one-to-one.
3   Find the equation of the inverse of a function.
4   Graph f^{-1} from the graph of f.

Key Terms

Use the vocabulary terms listed below to complete each statement in exercises 1−2.

> **one-to-one function**   **inverse of a function** f

1.   A function in which each x-value corresponds to just one y-value and each y-value corresponds to just one x-value is a(n) _____.

2.   If f is a one-to-one function, the _____ is the set of all ordered pairs of the form (y, x) where (x, y) belongs to f.

Objective 1   Decide whether a function is one-to-one and, if it is, find its inverse.

Video Examples

Review these examples for Objective 1:

1.   Find the inverse of each function that is one-to-one.

   a. $F = \{(2, 1), (-1, 1), (0, 0), (1, 1)\}$

   Every x-value in F corresponds to only one y-value. However, the y-value 1 corresponds to two x-values, so F is not a one-to-one function.

   b. $G = \{(-3,-1), (-2, 0), (-1, 1), (0, 2)\}$

   Every x-value in G corresponds to only one y-value, and every y-value corresponds to only one x-value, so G is a one-to-one function.
   The inverse function is found by interchanging the x- and y-values in each ordered pair.

   $G^{-1} = \{(-1,-3), (0,-2), (1,-1), (2, 0)\}$

Now Try:

1.   Find the inverse of each function that is one-to-one.

   a. $F = \{(2, 4), (-1, 1), (0, 0)$ $(1, 1), (2, 6)\}$

   _____

   b. $G = \{(3, 2), (-3,-2), (2, 3), (-2,-3)\}$

   _____

c.

State	Number of National Parks
AK	8
AZ	3
CA	8
CO	4
FL	3
HI	2
UT	5

c.

State	Number of representatives
AK	1
AZ	8
CA	53
FL	25
NY	29
DE	1

Let N be the function defined in the table, with the states forming the domain and the number of national parks forming the range. Then, N is not one-to-one, because two different states have the same number of national parks.

Objective 1 Practice Exercises

For extra help, see Example 1 on page 583 of your text.

If the function is one-to-one, find its inverse.

1. $\{(-3,-1), (-2, 2), (-1, 3), (0, 4)\}$ 1. _____

2. $\{(1, 0), (2, 0), (3, 5), (4, 1)\}$ 2. _____

3. $\{(0, 0), (1, 1), (-1, -1), (2, 2), (-2, -2)\}$ 3. _____

Objective 2   **Use the horizontal line test to determine whether a function is one-to-one.**

Video Examples

Review these examples for Objective 2:	**Now Try:**
2. Use the horizontal line test to determine whether each graph is the graph of a one-to-one function.	2. Use the horizontal line test to determine whether each graph is the graph of a one-to-one function.

a.

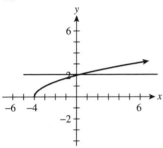

Every horizontal line will intersect the graph in exactly one point. The function is one-to-one.

b.

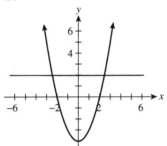

Because a horizontal line intersects the graph in more than one point, the function is not one-to-one.

a.

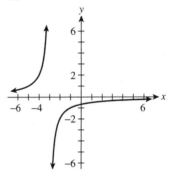

_____

b.

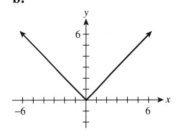

_____

Objective 2 Practice Exercises

For extra help, see Example 2 on page 584 of your text.

Use the horizontal line test to determine whether each function is one-to-one.

4.

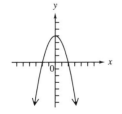

4. _____

5.

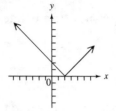

6.

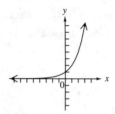

Objective 3   Find the equation of the inverse of a function.

Video Examples

Review these examples for Objective 3:

3. Decide whether each equation represents a one-to-one function. If so, find the equation for the inverse.

a. $f(x) = 3x - 5$

The graph of $y = 3x - 5$ is a nonvertical line, so by the horizontal line test, f is a one-to-one function. To find the inverse, let $y = f(x)$, interchange x and y, then solve for y.

$$y = 3x - 5$$

$$x = 3y - 5 \quad \text{Interchange } x \text{ and } y.$$

$$x + 5 = 3y$$

$$\frac{x + 5}{3} = y$$

$$f^{-1}(x) = \frac{x + 5}{3} = \frac{x}{3} + \frac{5}{3}$$

$$f^{-1}(x) = \frac{1}{3}x + \frac{5}{3}$$

b. $f(x) = 2x^2 + 3$

The graph of $y = 2x^2 + 3$ is a vertical parabola, so by the horizontal line test, f is not a one-to-one function and does not have an inverse.

Now Try:

3. Decide whether each equation represents a one-to-one function. If so, find the equation for the inverse.

a. $f(x) = 4x - 1$

_____

b. $f(x) = -\frac{3}{2}x^2$

_____

c. $f(x) = x^3 + 1$ **c.** $f(x) = 2x^3 - 3$

The graph of $y = x^3 + 1$ is a cubing function.
The function is one-to-one and has an inverse.

$$y = x^3 + 1$$

$$x = y^3 + 1 \quad \text{Interchange } x \text{ and } y.$$

$$x - 1 = y^3$$

$$\sqrt[3]{x - 1} = y$$

$$f^{-1}(x) = \sqrt[3]{x - 1}$$

_____

Objective 3 Practice Exercises

For extra help, see Examples 3–4 on pages 584–586 of your text.

If the function is one-to-one, find its inverse.

7. $f(x) = 2x - 5$ 7. _____

8. $f(x) = x^3 - 1$ 8. _____

9. $f(x) = x^2 - 1$ 9. _____

Objective 4   Graph f^{-1} from the graph of f.

Video Examples

Review this example for Objective 4:
5. Use the given graph to graph the inverse of f.

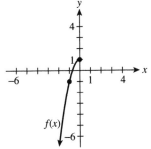

We can find the graph of f^{-1} from the graph of f
by locating the mirror image of each point in f

Now Try:
5. Use the given graph to graph the inverse of f.

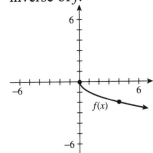

with respect to the line $y = x$.

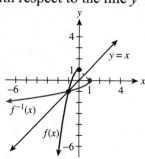

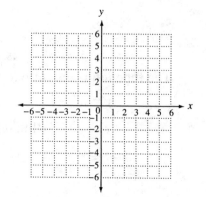

Objective 4 Practice Exercises

For extra help, see Example 5 on pages 586–587 of your text.

If the function is one-to-one, graph the function f and its inverse f^{-1} on the same set of axes.

10.

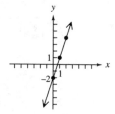

10. _____

11.

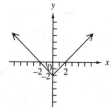

11. _____

12.

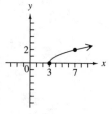

12. _____

Chapter 9 INVERSE, EXPONENTIAL, AND LOGARITHMIC FUNCTIONS

9.2 Exponential Functions

Learning Objectives
1   Use a calculator to find approximations of exponentials.
2   Define and graph exponential functions.
3   Solve exponential equations of the form $a^x = a^k$ for x.
4   Use exponential functions in applications involving growth or decay.

Key Terms

Use the vocabulary terms listed below to complete each statement in exercises 1−2.

exponential equation   **inverse**

1. If f is a one-to-one function, then the _____ of f is the set of all ordered pairs formed by interchanging the coordinates of the ordered pairs of f.

2. An equation that has a variable as an exponent, is an _____.

Objective 1   Use a calculator to find approximations of exponentials.

Video Examples

Review these examples for Objective 1:

1. Use a calculator to find an approximation to three decimal places for each exponential expression.

     a. $3^{1.8}$

     $3^{1.8} \approx 7.225$

     b. $3^{-1.4}$

     $3^{-1.4} \approx 0.215$

     c. $3^{1/4}$

     $3^{1/4} \approx 1.316$

Now Try:

1. Use a calculator to find an approximation to three decimal places for each exponential expression.

     a. $3^{1.9}$

     _____

     b. $3^{-1.6}$

     _____

     c. $3^{1/5}$

     _____

Name: Date:

Instructor: Section:

Objective 1 Practice Exercises

For extra help, see Example 1 on page 591 of your text.

Use a calculator to find an approximation to three decimal places for each exponential expression.

1. $3^{1.2}$

2. $3^{-1.2}$

3. $3^{1/3}$

1. _____

2. _____

3. _____

Objective 2   Define and graph exponential functions.

Video Examples

Review these examples for Objective 2:

2.   Graph $f(x) = 6^x$.

Create a table of values, then plot the points and draw a smooth curve through them.

x	$f(x) = 6^x$
-2	$\dfrac{1}{36}$
-1	$\dfrac{1}{6}$
0	1
1	6
2	36
3	216

3.   Graph $f(x) = \left(\dfrac{1}{6}\right)^x$.

Create a table of values, then plot the points and draw a smooth curve through them.

x	$f(x) = \left(\dfrac{1}{6}\right)^x$
-3	216
-2	36
-1	6
0	1
1	$\dfrac{1}{6}$
2	$\dfrac{1}{36}$

Now Try:

2.   Graph $f(x) = 3^x$.

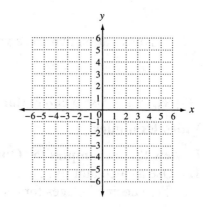

3.   Graph $f(x) = \left(\dfrac{1}{3}\right)^x$.

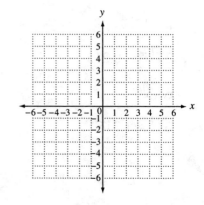

       Copyright © 2016 Pearson Education, Inc.

4. Graph $f(x) = 3^{2x-1}$.

Create a table of values, then plot the points and draw a smooth curve through them.

x	$2x-1$	$f(x) = 3^{2x-1}$
-1	-3	$\dfrac{1}{27}$
0	-1	$\dfrac{1}{3}$
1	1	3
2	3	27

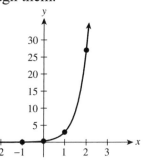

4. Graph $f(x) = 2^{1-x}$.

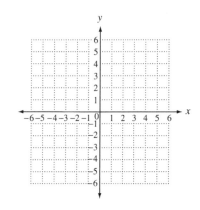

Objective 2 Practice Exercises

For extra help, see Examples 2–4 on pages 592–593 of your text.

Graph each exponential function.

4. $f(x) = 2^{-x}$

4.

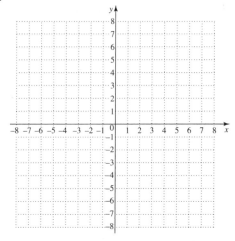

5. $f(x) = \left(\dfrac{1}{8}\right)^{x}$

5.

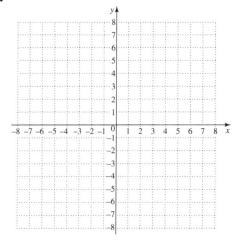

6. $f(x) = 4^{2x-3}$

6.

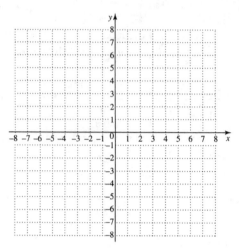

Objective 3 Solve exponential equations of the form $a^x = a^k$ for x.

Video Examples

Review these examples for Objective 3:

5. Solve the equation $16^x = 64$.

$$16^x = 64$$

$(2^4)^x = 2^6$ Write with the same base.

$2^{4x} = 2^6$ Power rule for exponents

$4x = 6$ If $a^x = a^y$, then $x = y$.

$x = \dfrac{6}{4} = \dfrac{3}{2}$ Solve for x; simplify.

Check: Substitute 3/2 for x.

$$16^{3/2} = (16^{1/2})^3 = 4^3 = 64$$

The solution set is $\left\{\dfrac{3}{2}\right\}$.

6. Solve each equation.

a. $16^{x-2} = 64^x$

$$16^{x-2} = 64^x$$

$(2^4)^{x-2} = (2^6)^x$ Write with the same base.

$2^{4x-8} = 2^{6x}$ Power rule for exponents

$4x - 8 = 6x$ If $a^x = a^y$, then $x = y$.

$-8 = 2x$ Solve for x.

$-4 = x$

The solution set is $\{-4\}$.

Now Try:

5. Solve the equation $25^x = 125$.

_____

6. Solve each equation.

a. $4^{x-1} = 8^x$

_____

b. $4^x = \dfrac{1}{64}$

$4^x = \dfrac{1}{64}$

$4^x = \dfrac{1}{4^3}$ $64 = 4^3$

$4^x = 4^{-3}$ Write with the same base.

$x = -3$ Set exponents equal.

The solution set is $\{-3\}$.

c. $\left(\dfrac{2}{5}\right)^x = \dfrac{125}{8}$

$\left(\dfrac{2}{5}\right)^x = \dfrac{125}{8}$

$\left(\dfrac{2}{5}\right)^x = \left(\dfrac{8}{125}\right)^{-1}$

$\left(\dfrac{2}{5}\right)^x = \left[\left(\dfrac{2}{5}\right)^3\right]^{-1}$ Write with the same base.

$\left(\dfrac{2}{5}\right)^x = \left(\dfrac{2}{5}\right)^{-3}$ Power rule for exponents

$x = -3$ Set exponents equal.

The solution set is $\{-3\}$.

b. $3^x = \dfrac{1}{243}$

_____

c. $\left(\dfrac{3}{2}\right)^x = \dfrac{16}{81}$

_____

Objective 3 Practice Exercises

For extra help, see Examples 5–6 on pages 594–595 of your text.

Solve each equation.

7. $25^{1-t} = 5$

7. _____

8. $8^{2x+1} = 4^{4x}$

8. _____

9. $\left(\dfrac{3}{4}\right)^x = \dfrac{16}{9}$

9. _____

337

Objective 4 Practice Exercises

Video Examples

Review these examples for Objective 4:

7. Suppose the number of bacteria present in a certain culture after t minutes is given by the equation $Q(t) = 2500(2^{0.05t})$,

How many bacteria were present after 20 minutes?

Start with the given function. Replace t with 20.

$$Q(t) = 2500(2^{0.05t})$$

$$Q(20) = 2500(2^{0.05 \times 20})$$

$$Q(20) = 2500(2^1)$$

$$Q(20) = 5000$$

There were 5000 bacteria present after 20 minutes.

8. The amount of radioactive material in a sample is given by the function $A(t) = 90\left(\dfrac{1}{2}\right)^{t/18}$, where $A(t)$ is the amount present, in grams, t days after the initial measurement.

How many grams will be present after 3 days? Round to the nearest hundredth.

Start with the given function. Replace t with 3.

$$A(t) = 90\left(\frac{1}{2}\right)^{t/18}$$

$$A(3) = 90\left(\frac{1}{2}\right)^{3/18}$$

$$A(3) \approx 80.18$$

After 3 days, there were about 80.18 grams in the sample.

Now Try:

7. The population of Evergreen Park is now 16,000. The population t years from now is given by the formula

$$P = 16,000(2^{t/10}).$$

Using the model, what will be the population 40 years from now?

_____

8. An industrial city in Ohio has found that its population is declining according to the equation $y = 70,000(2)^{-0.01x}$, where x is the time in years from 1910.

According to the model, what will the city's population be in the year 2020?

_____

Objective 4   Use exponential functions in applications involving growth or decay.

For extra help, see Examples 7–8 on pages 595–596 of your text.

Solve each problem.

10. The population of Canadian geese that spend the   10. _____
     summer at Gemini Lake each year has been growing
     according to the function $f(x) = 56(2)^{0.2x}$, where x
     is the time in years from 1990. Find the number of
     geese in 2010.

11. A sample of a radioactive substance with mass in   11. _____
     grams decays according to the function
     $f(x) = 100(10)^{-0.2x}$, where x is the time in hours
     after the original measurement. Find the mass of the
     substance after 10 hours.

12. A culture of a certain kind of bacteria grows   12. _____
     according to $f(x) = 7750(x)^{0.75x}$, where x is the
     number of hours after 12 noon. Find the number of
     bacteria in the culture at 12 noon.

Chapter 9   INVERSE, EXPONENTIAL, AND LOGARITHMIC FUNCTIONS

9.3   Logarithmic Functions

Learning Objectives

1   Define a logarithm.
2   Convert between exponential and logarithmic forms, and evaluate logarithms.
3   Solve logarithmic equations of the form $\log_a b = k$ for a, b, or k.
4   Use the definition of logarithm to simplify logarithmic expressions.
5   Define and graph logarithmic functions.
6   Use logarithmic functions in applications involving growth or decay.

Key Terms

Use the vocabulary terms listed below to complete each statement in exercises 1−2.

logarithm   **logarithmic equation**

1.   The _____ of a positive number is the exponent indicating the power to which it is necessary to raise a given number (the base) to give the original number.

2.   An equation with a logarithm in at least one term is a _____.

Objective 1   Define a logarithm.

For extra help, see pages 599–600 of your text.

Objective 2   Convert between exponential and logarithmic forms, and evaluate logarithms.

Video Examples

Review these examples for Objective 2:	Now Try:
1.	**1.**
a. Write $5^3 = 125$ in logarithmic form.	**a.** Write $8^2 = 64$ in logarithmic form.
$\log_5 125 = 3$	
	_____
b. Write $\log_{16} 4 = \dfrac{1}{2}$ in exponential form.	**b.** Write $\log_{16} \dfrac{1}{4} = -\dfrac{1}{2}$ in exponential form.
$16^{1/2} = 4$	
	_____

2. Use a calculator to approximate each logarithm to four decimal places.

   a. $\log_2 9$

   $\log_2 9 \approx 3.1699$

   b. $\log_7 15$

   $\log_7 15 \approx 1.3917$

   c. $\log_{1/3} 20$

   $\log_{1/3} 20 \approx -2.7268$

   d. $\log_{10} 17$

   $\log_{10} 17 \approx 1.2304$

2. Use a calculator to approximate each logarithm to four decimal places.

   a. $\log_2 6$

   _____

   b. $\log_4 21$

   _____

   c. $\log_{1/4} 25$

   _____

   d. $\log_{10} 12$

   _____

Objective 2 Practice Exercises

For extra help, see Examples 1–2 on pages 600–601 of your text.

Write in exponential form.

   1. $\log_{10} 0.001 = -3$

   1. _____

Write in logarithmic form.

   2. $2^{-7} = \dfrac{1}{128}$

   2. _____

Use a calculator to approximate the logarithm to four decimal places.

   3. $\log_3 25$

   3. _____

Objective 3   Solve logarithmic equations of the form $\log_a b = k$ for a, b, or k.

Video Examples

Review these examples for Objective 3:

3. Solve each equation.

   a. $\log_{3/2} x = -2$

   By definition, $\log_{3/2} x = -2$ is equivalent to

   $x = \left(\dfrac{3}{2}\right)^{-2}$, and $\left(\dfrac{3}{2}\right)^{-2} = \left(\dfrac{2}{3}\right)^{2} = \dfrac{4}{9}$. The solution

   set is $\left\{\dfrac{4}{9}\right\}$.

Now Try:

2. Solve each equation.

   a. $\log_4 x = -3$

   _____

b. $\log_5(3x+1)=2$ **b.** $\log_9(2x+1)=2$

$\log_5(3x+1)=2$

$\quad 3x+1=5^2$ Write in exponential form.

$\quad\quad 3x=24$ Apply the exponent; subtract 1.

$\quad\quad\quad x=8$ Divide by 3.

The solution set is $\{8\}$.

_____

c. $\log_x 6=2$ **c.** $\log_x 12=2$

$\log_x 6=2$

$\quad x^2=6$ Write in exponential form.

$\quad x=\pm\sqrt{6}$ Take square root.

Only the principal square root satisfies the equation since the base must be a positive number. The solution set is $\left\{\sqrt{6}\right\}$.

_____

d. $\log_{64}\sqrt[4]{8}=x$ **d.** $\log_{81}\sqrt[3]{9}=x$

$\log_{64}\sqrt[4]{8}=x$

$\quad\quad 64^x=\sqrt[4]{8}$ Write in exponential form.

$\quad\left(8^2\right)^x=8^{1/4}$ Write with the same base.

$\quad\quad 8^{2x}=8^{1/4}$ Power rule for exponents

$\quad\quad 2x=\dfrac{1}{4}$

$\quad\quad\quad x=\dfrac{1}{8}$

The solution set is $\left\{\dfrac{1}{8}\right\}$.

_____

Objective 3 Practice Exercises

For extra help, see Example 3 on pages 601–602 of your text.

Solve each equation.

4. $x=\log_{32}8$ **4.** _____

5. $\log_{1/3}r=-4$ **5.** _____

6. $\log_a 4 = \dfrac{1}{2}$

6. _____

Objective 4   Use the definition of logarithm to simplify logarithmic expressions.

Video Examples

Review these examples for Objective 4:

4. Use special properties to evaluate each expression.

   a. $\log_8 8$

   $\log_8 8 = 1$

   b. $\log_{64} 1$

   $\log_{64} 1 = 0$

   c. $\log_{0.2} 1$

   $\log_{0.2} 1 = 0$

   d. $\log_4 4^{11}$

   $\log_4 4^{11} = 11$

   e. $8^{\log_8 5}$

   $8^{\log_8 5} = 5$

   f. $\log_4 64$

   $\log_4 64 = \log_4 4^3 = 3$

Now Try:

4. Use special properties to evaluate each expression.

   a. $\log_4 4$

   _____

   b. $\log_{100} 1$

   _____

   c. $\log_{1/3} 1$

   _____

   d. $\log_6 6^3$

   _____

   e. $6^{\log_6 9}$

   _____

   f. $\log_5 125$

   _____

Objective 4 Practice Exercises

For extra help, see Example 4 on page 602 of your text.

Use the special properties to evaluate each expression.

7. $\log_{3.4} 1$

7. _____

8. $\log_8 8^3$

8. _____

9. $2^{\log_2 5}$

9. _____

Objective 5   Define and graph logarithmic functions.

Video Examples

Review these examples for Objective 5:

5. Graph $f(x) = \log_5 x$.

Begin by writing $y = \log_5 x$ in exponential form as $x = 5^y$. Then, create a table of values, plot the points and draw a smooth curve through them.

$x = 5^y$	y
$\frac{1}{5}$	-1
1	0
5	1
25	2

Now Try:

5. Graph $f(x) = \log_3 x$.

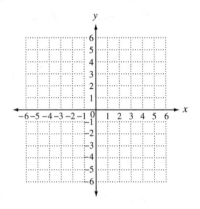

6. Graph $f(x) = \log_{1/3} x$.

Begin by writing $y = \log_{1/3} x$. in exponential form. Then, create a table of values, plot the points and draw a smooth curve through them.

$x = \left(\frac{1}{3}\right)^y$	y
$\frac{1}{3}$	1
1	0
3	-1
9	-2

6. Graph $f(x) = \log_{1/4} x$.

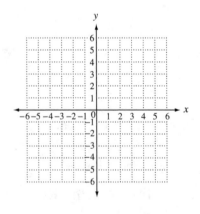

Objective 5 Practice Exercises

For extra help, see Examples 5–6 on page 603 of your text.

Graph each logarithmic function.

10. $y = \log_9 x$

10.

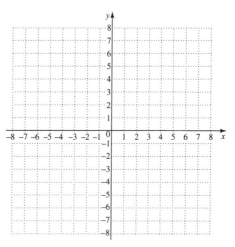

11. $y = \log_{1/4} x$

11.

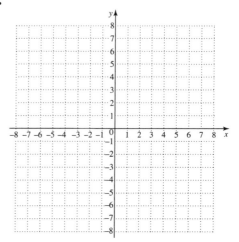

Objective 6   Use logarithmic functions in applications involving growth or decay.

Video Examples

Review this example for Objective 6:

6. A company analyst has found that total sales in thousands of dollars after a major advertising campaign are given by $S(x) = 100 \log_2 (x+2)$, where x is time in weeks after the campaign was introduced. Find the amount of sales two weeks after the campaign was introduced.

Two weeks after the campaign, $x = 2$, so we have
$$S(2) = 100 \log_2 (2+2)$$
$$S(2) = 100 \log_2 (2)^2$$
$$S(2) = 100 \cdot 2$$
$$S(2) = 200$$

Two weeks after the campaign, sales were $200,000.

Now Try:

6. The number of fish in an aquarium is given by the function $f(t) = 8 \log_5 (2t + 5)$, where t is time in months. Find the number of fish present after 10 months.

_____

Objective 6 Practice Exercises

For extra help, see Example 7 on page 604 of your text.

Solve each problem.

12. The population of foxes in an area t months after the foxes were introduced there is approximated by the function $F(t) = 500 \log_{10}(2t+10)$. Find the number of foxes in the area when the foxes were first introduced into the area.

12. _____

13. A population of mites in a laboratory is growing according to the function
$p = 50 \log_3 (20t + 7) - 25 \log_9 (80t + 1)$, where t is the number of days after a study is begun. Find the number of mites present 1 day after the beginning of the study.

13. _____

14.   Sales (in thousands) of a new product are   14. _____
      approximated by
      $S = 125 + 20\log_2(30t + 4) + 30\log_4(35t - 6),$ where
      t is the number of years after the product is
      introduced. Find the total sales 2 years after the
      product is introduced.

Chapter 9   INVERSE, EXPONENTIAL, AND LOGARITHMIC FUNCTIONS

9.4   Properties of Logarithms

Learning Objectives
1   Use the product rule for logarithms.
2   Use the quotient rule for logarithms.
3   Use the power rule for logarithms.
4   Use properties to write alternative forms of logarithmic expressions.

Key Terms

Use the vocabulary terms listed below to complete each statement in exercises 1–4.

> **product rule for logarithms**   **quotient rule for logarithms**
>
> **power rule for logarithms**   **special properties**

1. The equations $b^{\log_b x} = x$, $x > 0$ and $\log_b b^x = x$ are referred to as the

   _____ of logarithms.

2. The equation $\log_b \frac{x}{y} = \log_b x - \log_b y$ is referred to as the _____.

3. The equation $\log_b xy = \log_b x + \log_b y$ is referred to as the _____.

4. The equation $\log_b x^r = r \log_b x$ is referred to as the _____.

Objective 1   Use the product rule for logarithms.

Video Examples

Review these examples for Objective 1:

1. Use the product rule to rewrite each logarithm. Assume $x > 0$.

   a. $\log_4 (6 \cdot 11)$

   Use the product rule.
   $\log_4 (6 \cdot 11) = \log_4 6 + \log_4 11$

   b. $\log_3 8 + \log_3 2$

   Use the product rule.
   $\log_3 8 + \log_3 2 = \log_3 (8 \cdot 2) = \log_3 16$

   c. $\log_2 (2x)$

   $\log_2 (2x) = \log_2 2 + \log_2 x$ Product rule
   $\qquad\quad = 1 + \log_2 x$ $\log_2 2 = 1$

Now Try:

1. Use the product rule to rewrite each logarithm. Assume $x > 0$.

   a. $\log_6 (5 \cdot 3)$

   _____

   b. $\log_5 7 + \log_5 3$

   _____

   c. $\log_4 (4x)$

   _____

d. $\log_3 x^4$

$$\log_3 x^4 = \log_3 (x \cdot x \cdot x \cdot x)$$
$$= \log_3 x + \log_3 x + \log_3 x + \log_3 x$$
$$= 4 \log_3 x$$

d. $\log_5 x^4$

_____

Objective 1 Practice Exercises

For extra help, see Example 1 on page 608 of your text.

Use the product rule to express each logarithm as a sum of logarithms.

1. $\log_7 5m$

2. $\log_2 6xy$

1. _____

2. _____

Use the product rule to express the sum as a single logarithm.

3. $\log_4 7 + \log_4 3$

3. _____

Objective 2   Use the quotient rule for logarithms.

Video Examples

Review these examples for Objective 2:

2.   Use the quotient rule to rewrite each logarithm. Assume $x > 0$.

a. $\log_4 \dfrac{8}{7}$

$$\log_4 \frac{8}{7} = \log_4 8 - \log_4 7$$

b. $\log_3 8 - \log_3 x$

$$\log_3 8 - \log_3 x = \log_3 \frac{8}{x}$$

c. $\log_2 \dfrac{16}{11}$

$$\log_2 \frac{16}{11} = \log_2 16 - \log_2 11 = 4 - \log_2 11$$

Now Try:

2.   Use the quotient rule to rewrite each logarithm. Assume $x > 0$.

a. $\log_5 \dfrac{4}{9}$

_____

b. $\log_6 x - \log_6 3$

_____

c. $\log_4 \dfrac{16}{11}$

_____

Objective 2 Practice Exercises

For extra help, see Example 2 on page 609 of your text.

Use the quotient rule for logarithms to express each logarithm as a difference of logarithms, or as a single number if possible.

4. $\log_2 \dfrac{5}{m}$

4. _____

5. $\log_6 \dfrac{k}{3}$

5. _____

Use the quotient rule for logarithms to express the difference as a single logarithm.

6. $\log_2 7q^4 - \log_2 5q^2$

6. _____

Objective 3   Use the power rule for logarithms.

Video Examples

Review these examples for Objective 3:

3.   Use the power rule to rewrite each logarithm. Assume that $b > 0$, $x > 0$, and $b \neq 1$.

a. $\log_4 3^5$

$\log_4 3^5 = 5\log_4 3$

b. $\log_b \sqrt{11}$

$\log_b \sqrt{11} = \log_b 11^{1/2} = \dfrac{1}{2}\log_b 11$

c. $\log_2 \sqrt[5]{x^4}$

$\log_2 \sqrt[5]{x^4} = \log_2 x^{4/5} = \dfrac{4}{5}\log_2 x$

d. $\log_3 \dfrac{1}{x^5}$

$\log_3 \dfrac{1}{x^5} = \log_3 x^{-5} = -5\log_3 x$

Now Try:

3.   Use the power rule to rewrite each logarithm. Assume that $b > 0$, $x > 0$, and $b \neq 1$.

a. $\log_6 4^3$

_____

b. $\log_b \sqrt{13}$

_____

c. $\log_3 \sqrt[4]{x^3}$

_____

d. $\log_3 \dfrac{1}{x^7}$

_____

Objective 3 Practice Exercises

For extra help, see Example 3 on page 610 of your text.

Use the power rule for logarithms to rewrite each logarithm or as a single number if possible.

7. $\log_m 2^7$ 7. _____

8. $\log_3 \sqrt[3]{5}$ 8. _____

9. $3^{\log_3 \sqrt[3]{7}}$ 9. _____

Objective 4   Use properties to write alternative forms of logarithmic expressions.

Video Examples

Review these examples for Objective 4:

4. Use the properties of logarithms to rewrite each expression if possible. Assume that all variables represent positive real numbers.

   a. $\log_5 6x^3$

   $$\log_5 6x^3 = \log_5 6 + \log_5 x^3$$
   $$= \log_5 6 + 3\log_5 x$$

   b. $\log_b \sqrt{\dfrac{5}{x}}$

   $$\log_b \sqrt{\dfrac{5}{x}} = \log_b \left(\dfrac{5}{x}\right)^{1/2}$$
   $$= \dfrac{1}{2}\log_b \dfrac{5}{x}$$
   $$= \dfrac{1}{2}(\log_b 5 - \log_b x)$$

Now Try:

4. Use the properties of logarithms to rewrite each expression if possible. Assume that all variables represent positive real numbers.

   a. $\log_6 36x^3$

   _____

   b. $\log_b \sqrt{\dfrac{x}{3}}$

   _____

c. $3\log_b x - \left(2\log_b y + \dfrac{1}{2}\log_b z\right)$

$3\log_b x - \left(2\log_b y + \dfrac{1}{2}\log_b z\right)$

$= \log_b x^3 - (\log_b y^2 + \log_b z^{1/2})$

$= \log_b x^3 - \log_b y^2 \sqrt{z}$

$= \log_b \dfrac{x^3}{y^2 \sqrt{z}}$

d. $2\log_3 x + \log_3(x-1) - \dfrac{1}{2}\log_3(x+1)$

$2\log_3 x + \log_3(x-1) - \dfrac{1}{2}\log_3(x+1)$

$= \log_3 x^2 + \log_3(x-1) - \log_3(x+1)^{1/2}$

$= \log_3\left(x^2(x-1)\right) - \log_3\sqrt{x+1}$

$= \log_3 \dfrac{x^3 - x^2}{\sqrt{x+1}}$

e. $\log_2(2x+3y)$

$\log_2(2x+3y)$ cannot be rewritten using the properties of logarithms.

5. Given that $\log_2 9 \approx 3.1699$ and $\log_2 11 \approx 3.4594$, use properties of logarithms to evaluate each expression.

a. $\log_2 99$

$\log_2 99 = \log_2(9\cdot 11)$

$\qquad = \log_2 9 + \log_2 11$

$\qquad = 3.1699 + 3.4594$

$\qquad = 6.6293$

b. $\log_2 \dfrac{1}{9}$

$\log_2 \dfrac{1}{9} = \log_2 1 - \log_2 9$

$\qquad = 0 - 3.1699$

$\qquad = -3.1699$

c. $\log_b x + 4\log_b y - \log_b z$

_____

d.

$2\log_2 x + \log_2(x-1) - \dfrac{1}{3}\log_2(x^2+1)$

_____

e. $\log_3(3x-y)$

_____

5. Given that $\log_2 9 \approx 3.1699$ and $\log_2 11 \approx 3.4594$, use properties of logarithms to evaluate each expression.

a. $\log_2 198$

_____

b. $\log_2 \dfrac{1}{11}$

_____

c. $\log_2 121$

$$\log_2 121 = \log_2 11^2$$
$$= 2\log_2 11$$
$$= 2(3.4594)$$
$$= 6.9188$$

6. Decide whether each statement is *true* or *false*.

a. $\log_2 32 - \log_2 16 = \log_2 16$

Evaluate each side.
Left side:
$$\log_2 32 - \log_2 16 = \log_2 2^5 - \log_2 2^4$$
$$= 5 - 4\log_2 2$$
$$= 1$$
Right side:
$$\log_2 16 = \log_2 2^4 = 4$$
The statement is false because $1 \neq 4$.

b. $\log_3 (\log_4 64) = \dfrac{\log_{12} 144}{\log_6 36}$

Evaluate each side.
Left side:
$$\log_3 (\log_4 64) = \log_3 (\log_4 4^3)$$
$$= \log_3 3$$
$$= 1$$
Right side:
$$\frac{\log_{12} 144}{\log_6 36} = \frac{\log_{12} 12^2}{\log_6 6^2} = \frac{2}{2} = 1$$
The statement is true because $1 = 1$.

c. $\log_2 729$

_____

6. Decide whether each statement is *true* or *false*.

a. $\log_3 27 + \log_3 9 = \log_5 5$

_____

b. $(\log_2 8)(\log_2 4) = \log_2 32$

_____

For extra help, see Examples 4–6 on pages 611–612 of your text.

Use the properties of logarithms to express the sum or difference of logarithms as a single logarithm, or as a single number if possible.

10. $\log_4 10y + \log_4 3y - \log_4 6y^3$

10. _____

Given that $\log_2 6 \approx 2.5850$ and $\log_2 12 \approx 3.5850$, *use properties of logarithms to evaluate the expression.*

11. $\log_2 72$

11. _____

Decide whether the statement is true *or* false.

12. $\log_2 4p^3 = 6 + 3\log_2 p$

12. _____

Chapter 9 INVERSE, EXPONENTIAL, AND LOGARITHMIC FUNCTIONS

9.5   Common and Natural Logarithms

Learning Objectives
1   Evaluate common logarithms using a calculator.
2   Use common logarithms in applications.
3   Evaluate natural logarithms using a calculator.
4   Use natural logarithms in applications.
5   Use the change-of-base rule.

Key Terms

Use the vocabulary terms listed below to complete each statement in exercises 1–2.

common logarithm   **natural logarithm**

1.   A logarithm to the base e is a _____.

2.   A logarithm to the base 10 is a _____.

Objective 1   Evaluate common logarithms using a calculator.

Video Examples

Review this example for Objective 1:

1.   Evaluate the logarithm to four decimal places using a calculator.

$\log 436.2$

$\log 436.2 \approx 2.6397$

Now Try:

1.   Evaluate the logarithm to four decimal places using a calculator.
$\log 983.5$

_____

Objective 1 Practice Exercises

For extra help, see Example 1 on pages 614–615 of your text.

Use a calculator to find each logarithm. Give an approximation to four decimal places.

1.   log 57.23   1. _____

2.   log 0.0914   2. _____

3.   log 87,123   3. _____

Objective 2   Use common logarithms in applications.

Video Examples

Review these examples for Objective 2:

2. Wetlands are classified as bogs, fens, marshes, and swamps, on the basis of pH values. A pH value between 6.0 and 7.5 indicates that the wetland is a "rich fen." When the pH is between 3.0 and 6.0, the wetland is a "poor fen," and if the pH falls to 3.0 or less, it is a "bog." Suppose that the hydronium ion concentration of a sample of water from a wetland is 5.4×10^{-4}. Find the pH value for the water and determine how the wetland should be classified.

$$pH = -\log\left(5.4 \times 10^{-4}\right) \quad \text{Definition of pH}$$

$$= -\left(\log 5.4 + \log 10^{-4}\right) \quad \text{Product rule}$$

$$= -\left(0.7324 - 4\right) \quad \begin{array}{l}\text{Use a calculator to} \\ \text{find } \log 5.4.\end{array}$$

$$= 3.2676$$

Since the pH is between 3.0 and 6.0, the wetland is a poor fen.

3. Find the hydronium ion concentration of a solution with pH 5.4.

$$pH = -\log\left[H_3O^+\right] \quad \text{Definition of pH}$$

$$5.4 = -\log\left[H_3O^+\right]$$

$$\log\left[H_3O^+\right] = -5.4 \qquad \text{Multiply by } -1.$$

$$H_3O^+ = 10^{-5.4} \qquad \begin{array}{l}\text{Write in} \\ \text{exponential form.}\end{array}$$

$$\approx 4.0 \times 10^{-6} \quad \text{Use a calculator.}$$

4. Find the decibel level to the nearest whole number of the sound with intensity I of $5.012 \times 10^{10} I_0$.

$$D = 10\log\left(\frac{I}{I_0}\right) = 10\log\left(\frac{5.012 \times 10^{10} I_0}{I_0}\right)$$

$$= 10\log\left(5.012 \times 10^{10}\right)$$

$$\approx 107 \text{ db}$$

Now Try:

2. Suppose that the hydronium ion concentration of a sample of water from a wetland is 6.2×10^{-8}. Find the pH value for the water and determine how the wetland should be classified.

_____

3. Find the hydronium ion concentration of a solution with pH 3.6.

_____

4. Find the decibel level to the nearest whole number of the sound with intensity I of $3.16 \times 10^8 I_0$.

_____

Objective 2 Practice Exercises

For extra help, see Examples 2–4 on pages 615–616 of your text.

Solve each problem.

4. Find the pH of a solution with the given hydronium
   ion concentration. Round the answer to the nearest
   tenth.

   a. 4.3×10^{-9} **b.** 2.8×10^{-6}

   4. a. _____

   b. _____

5. Find the decibel level to the nearest whole number of
   the sound with intensity I of $2.5 \times 10^{13} I_0$.

   5. _____

6. Find the hydronium ion concentration of a solution
   with the given pH value.
   a. 5.2   **b.** 1.3

   6. a. _____

   b. _____

Objective 3   Evaluate natural logarithms using a calculator.

Video Examples

Review this example for Objective 3:

5. Using a calculator, evaluate the logarithm to four
   decimal places.

   $\ln 436.2$

   $\ln 436.2 \approx 6.0781$

Now Try:

5. Using a calculator, evaluate the
   logarithm to four decimal
   places.
   $\ln 98$

   _____

Objective 3 Practice Exercises

For extra help, see Example 5 on page 618 of your text.

Find each natural logarithm. Give an approximation to four decimal places.

7. $\ln 76.3$

   7. _____

8. $\ln 0.102$

   8. _____

9. $\ln 50$

   9. _____

Name: Date:

Instructor: Section:

Objective 4   Use natural logarithms in applications.

Video Examples

Review this example for Objective 4:

6.   The time t in years for an investment increasing at a rate of r percent (in decimal form) to double is given by

$$t = \frac{\ln 2}{\ln(1+r)}.$$

This is called the doubling time. Find the doubling time to the nearest tenth for an investment at 4%.

$4\% = 0.04$, so $t = \dfrac{\ln 2}{\ln(1+0.04)} = \dfrac{\ln 2}{\ln 1.04} \approx 17.7$

The doubling time for the investment is about 17.7 years.

Now Try:

6.   Use the formula at the left to find the doubling time to the nearest tenth for an investment at 6%.

_____

Objective 4 Practice Exercises

For extra help, see Example 6 on page 618 of your text.

The time t in years for an amount increasing at a rate of r (in decimal form) to double (the doubling time) is given by $t = \dfrac{\ln 2}{\ln(1+r)}$. *Find the doubling time for an investment at the interest rate. Round to the nearest whole number.*

10.   3%

10. _____

The half-life of a radioactive substance is the time it takes for half of the material to decay. The amount A in pounds of substance remaining after t years is given by $\ln \dfrac{A}{C} = -\dfrac{t}{h} \ln 2$, *where C is the initial amount in pounds, and h is its half-life in years. Use the formula to solve the following problems. Round to the nearest whole number.*

11.   The half-life of radium-226 is 1620 years. How long, to the nearest year, will it take for 100 pounds to decay to 25 pounds?

11. _____

Newton's Law of Cooling describes the cooling of a warmer object to the cooler temperature of the surrounding environment. The formula can be given as

$t = \dfrac{1}{k}\ln\dfrac{T_s - T_1}{T_s - T_2}$, *where t is the elapsed time, T_1 is the initial temperature measurement of the object, T_2 is the second temperature measurement of the object, and T_s is the temperature of the surrounding environment. Use this formula to solve the problem. Round to the nearest tenth.*

12.   A corpse was discovered in a motel room at midnight and its temperature was 80°F. The temperature in the room was 60°F. Assuming that the person's temperature at the time of death was 98.6° F and using $k = 0.1438$, determine t and the time of death.

12. t_____

time _____

Objective 5   Use the change-of-base rule.

Video Examples

Review this example for Objective 5:

7.   Evaluate $\log_7 28$ to four decimal places.

$$\log_7 28 = \frac{\log 28}{\log 7} = 1.7124$$

Now Try:

7.   Evaluate $\log_5 180$ to four decimal places.

_____

Objective 5 Practice Exercises

For extra help, see Example 7 on page 619 of your text.

Use the change-of-base rule to find each logarithm. Give approximations to four decimal places.

13. $\log_{16} 27$

13. _____

14. $\log_6 0.25$

14. _____

15. $\log_{1/2} 5$

15. _____

359

Chapter 9   INVERSE, EXPONENTIAL, AND LOGARITHMIC FUNCTIONS

9.6   Exponential and Logarithmic Equations; Further Applications

Learning Objectives
1   Solve equations involving variables in the exponents.
2   Solve equations involving logarithms.
3   Solve applications involving compound interest.
4   Solve applications involving base e exponential growth and decay.

Key Terms

Use the vocabulary terms listed below to complete each statement in exercises 1−2.

compound interest   **continuous compounding**

1.   The formula for _____ is $A = Pe^{rt}$.

2.   The formula for _____ is $A = P\left(1 + \frac{r}{n}\right)^{nt}$.

Objective 1   Solve equations involving variables in the exponents.

Video Examples

Review these examples for Objective 1:

1.   Solve $4^x = 30$. Approximate the solution to three decimal places.

$$4^x = 30$$

$\log 4^x = \log 30$ If $x = y$, and $x > 0$, $y > 0$, then $\log_b x = \log_b y$.

$x \log 4 = \log 30$ Power rule

$x = \dfrac{\log 30}{\log 4}$ Divide by log 4.

$x \approx 2.453$ Use a calculator.

Check

$$4^x = 4^{2.453} \approx 30$$

The solution set is {2.453}.

Now Try:

1.   Solve $3^x = 15$. Approximate the solution to three decimal places.

_____

2. Solve $e^{0.005x} = 9$. Approximate the solution to three decimal places.

$$e^{0.005x} = 9$$

$$\ln e^{0.005x} = \ln 9 \qquad \text{If } x = y, \text{ and } x > 0,$$
$$\qquad\qquad\qquad y > 0, \text{ then } \ln x = \ln y.$$

$$0.005x \ln e = \ln 9 \qquad \text{Power rule}$$

$$0.005x = \ln 9 \qquad \ln e = 1$$

$$x = \frac{\ln 9}{0.005} \qquad \text{Divide by 0.005.}$$

$$x \approx 439.445 \text{ Use a calculator.}$$

The solution set is $\{439.445\}$.

2. Solve $e^{0.4x} = 15$. Approximate the solution to three decimal places.

_____

Objective 1 Practice Exercises

For extra help, see Examples 1–3 on pages 623–624 of your text.

Solve each equation. Give solutions to three decimal places.

1. $25^{x+2} = 125^{3-x}$

1. _____

2. $4^{x-1} = 3^{2x}$

2. _____

3. $e^{-0.45x} = 7$

3. _____

Objective 2   Solve equations involving logarithms.

Video Examples

Review these examples for Objective 2:

Now Try:

5. Solve $\log_3 (x-1)^3 = 5$. Give the exact solution.

$\log_3 (x-1)^3 = 5$

$(x-1)^3 = 3^5$ Write in exponential form.

$(x-1)^3 = 243$

$x-1 = \sqrt[3]{243}$ Take the cube root of each side.

$x-1 = 3\sqrt[3]{9}$ Simplify the cube root.

$x = 1 + 3\sqrt[3]{9}$ Add 1.

Check:

$\log_3 (x-1)^3 = 5$

$\log_3 \left(1 + 3\sqrt[3]{9} - 1\right)^3 \overset{?}{=} 5$

$\log_3 \left(\sqrt[3]{243}\right)^3 \overset{?}{=} 5$

$\log_3 (243) \overset{?}{=} 5$

$\log_3 3^5 \overset{?}{=} 5$

$5 = 5$

The solution set is $\left\{1 + 3\sqrt[3]{9}\right\}$.

5. Solve $\log_6 (x+1)^3 = 2$. Give the exact solution.

_____

6. Solve $\log_3 (5x+42) - \log_3 x = \log_3 26$.

$\log_3 (5x+42) - \log_3 x = \log_3 26$

$\log_3 \dfrac{5x+42}{x} = \log_3 26$

$\dfrac{5x+42}{x} = 26$ If $\log_b x = \log_b y$ then $x = y$.

$5x + 42 = 26x$ Multiply by x.

$42 = 21x$ Subtract $5x$.

$2 = x$ Divide by 21.

6. Solve
$\log_6 (2x+7) - \log_6 x = \log_6 16$.

_____

Check:

$$\log_3 (5x+42) - \log_3 x = \log_3 26$$

$$\log_3 (5\cdot 2 + 42) - \log_3 2 \overset{?}{=} \log_3 26$$

$$\log_3 52 - \log_3 2 \overset{?}{=} \log_3 26$$

$$\log_3 \tfrac{52}{2} \overset{?}{=} \log_3 26$$

$$\log_3 26 = \log_3 26$$

The solution set is $\{2\}$.

7. Solve $\log_2 (x+7) + \log_2 (x+3) = \log_2 77$.

$$\log_2 (x+7) + \log_2 (x+3) = \log_2 77$$

$$\log_2 [(x+7)(x+3)] = \log_2 77$$

                  Product rule

$$(x+7)(x+3) = 77$$

                  If $\log_b x = \log_b y$
                  then $x = y$.

$$x^2 + 10x + 21 = 77 \quad \text{Multiply.}$$

$$x^2 + 10x - 56 = 0 \quad \text{Subtract 77.}$$

$$(x-4)(x+14) = 0 \quad \text{Factor.}$$

$$x - 4 = 0 \quad \text{or} \quad x + 14 = 0$$

$$x = 4 \qquad\qquad x = -14$$

The value -14 must be rejected since it leads to the logarithm of a negative number in the original equation.

A check shows that the only solution is 4.

The solution set is $\{4\}$.

7. Solve.

$$\log_4 (4x-3) + \log_4 x = \log_4 (2x-1)$$

_____

Objective 2 Practice Exercises

For extra help, see Examples 4–7 on pages 624–626 of your text.

Solve each equation. Give exact solution.

4. $\log(-a) + \log 4 = \log(2a+5)$

4. _____

5. $\log_3(x^2 - 10) - \log_3 x = 1$ **5.** _____

6. $\ln(x+4) + \ln(x-2) = \ln 7$ **6.** _____

Objective 3   Solve applications involving compound interest.

Video Examples

Review these examples for Objective 3:

8. How much money will be in an account at the end of 5 years if $5000 is deposited at 4% compounded monthly?

Because interest is compounded monthly, $n = 12$. The other given values are $P = 5000$, $r = 0.04$, and $t = 5$.

$$A = P\left(1 + \frac{r}{n}\right)^{nt}$$

$$A = 5000\left(1 + \frac{0.04}{12}\right)^{12 \cdot 5}$$

$$A = 5000(1.0033)^{60}$$

$$A = 6104.98$$

There will be $6104.98 in the account at the end of 5 years.

9. Approximate the time it would take for money deposited in an account paying 5% interest compounded quarterly to double. Round to the nearest hundredth.

We want the number of years t for P dollars to grow to $2P$ dollars at a rate of 5% per year. In the compound interest formula, we substitute $2P$ for A, and let $r = 0.05$ and $n = 4$.

Now Try:

8. How much money will be in an account at the end of 5 years if $10,000 is deposited at 4% compounded quarterly?

_____

9. Approximate the time it would take for money deposited in an account paying 5% interest compounded monthly to double. Round to the nearest hundredth.

_____

$$2P = P\left(1 + \frac{0.05}{4}\right)^{4t}$$

$$2 = 1.0125^{4t}$$

$$\log 2 = \log 1.025^{4t}$$

$$\log 2 = 4t \log 1.0125$$

$$t = \frac{\log 2}{4 \log 1.0125}$$

$$t \approx 13.95$$

It will take about 13.95 years for the investment to double.

10. Suppose that $5000 is invested at 4% interest for 3 years.

a. How much will the investment be worth if it is compounded continuously?

$$A = Pe^{rt}$$

$$A = 5000e^{0.04 \cdot 3}$$

$$A = 5000e^{0.12}$$

$$A = 5637.48$$

The investment will be worth $5637.48.

b. Approximate the amount of time it would take for the investment to double. Round to the nearest tenth.

Find the value of t that will cause A to be $2(\$5000) = \$10,000$.

$$A = Pe^{rt}$$

$$10,000 = 5000e^{0.04t}$$

$$2 = e^{0.04t} \qquad \text{Divide by 5000.}$$

$$\ln 2 = \ln e^{0.04t} \qquad \begin{array}{l}\text{If } x = y, \\ \text{then } \ln x = \ln y.\end{array}$$

$$\ln 2 = 0.04t \qquad \ln e^{k} = k$$

$$\frac{\ln 2}{0.04} = t \qquad \text{Divide by 0.04.}$$

$$t \approx 17.3$$

It will take about 17.3 years for the amount to double.

10. Suppose that $5000 is invested at 2% interest for 3 years.

a. How much will the investment be worth if it is compounded continuously?

_____

b. Approximate the amount of time it would take for the investment to double. Round to the nearest tenth.

_____

Objective 3 Practice Exercises

For extra help, see Examples 8–10 on pages 627–628 of your text.

Solve each problem.

7. How much will be in an account after 10 years if   7. _____
    $25,000 is invested at 8% compounded quarterly?
    Round to the nearest cent.

8. How much will be in an account after 5 years if   8. _____
    $10,000 is invested at 4.5% compounded
    continuously? Round to the nearest cent.

9. How long will it take an investment to double if it is   9. _____
    placed in an account paying 9% interest
    compounded continuously? Round to the nearest
    tenth.

Objective 4   Solve applications involving base *e* exponential growth and decay.

Video Examples

Review these examples for Objective 4:

11. A sample of 500 g of lead-210 decays according
     to the function $y = y_0 e^{-0.032t}$, where t is the
     time in years, y is the amount of the sample
     at time t, and y_0 is the initial amount present
     at $t = 0$.

     a. How much lead will be left in the sample
     after 20 years? Round to the nearest tenth of a
     gram.

     Let $t = 20$ and $y_0 = 500$.

     $y = 500e^{-0.032 \cdot 20} \approx 263.6$

     There will be about 263.6 grams after 20 years.

Now Try:

11. Cesium-137, a radioactive
     isotope used in radiation
     therapy, decays according to
     the function $y = y_0 e^{-0.0231t}$,
     where t is the time in years and
     y_0 is the initial amount present
     at $t = 0$.

     a. If an initial sample contains
     36 mg of cesium-137, how
     much cesium-137 will be left in
     the sample after 50 years?
     Round to the nearest tenth.

     _____

b. Approximate the half-life of lead-210 to the nearest tenth.

Let $y = \frac{1}{2}(500) = 250$.

$$250 = 500e^{-0.032t}$$

$$0.5 = e^{-0.032t}$$

$$\ln 0.5 = \ln e^{-0.032t}$$

$$\ln 0.5 = -0.032t$$

$$t = \frac{\ln 0.5}{-0.032} \approx 21.7$$

The half-life of lead-210 is about 21.7 years.

b. Approximate the half-life of cesium-137 to the nearest tenth.

_____

Objective 4 Practice Exercises

For extra help, see Example 11 on page 629 of your text.

Solve each problem.

10.   Radioactive strontium decays according to the function $y = y_0 e^{-0.0239t}$, where t is the time in years. If an initial sample contains $y_0 = 15$ g of radioactive strontium, how many grams will be present after 25 years? Round to the nearest hundredth of a gram.

10.   _____

11.   How long will it take the initial sample of strontium in exercise 19 to decay to half of its original amount?

11.   _____

12. The concentration of a drug in a person's system decreases according to the function $C(t) = 2e^{-0.2t}$, where $C(t)$ is given in mg and t is in hours. How much of the drug will be in the person's system after one hour? Approximate answer to the nearest hundredth.

12. _____

Chapter R REVIEW OF THE REAL NUMBER SYSTEM

R.1 Basic Concepts

Key Terms

1. number line
2. Set-builder notation
3. coordinate
4. set
5. finite set
6. empty set
7. inequality
8. signed numbers
9. infinite set
10. variable
11. absolute value
12. additive inverse
13. equation
14. graph
15. elements

Objective 1

Now Try

1. $\{0, 1, 2, 3, 4, 5\}$
2. $\{x \mid x \text{ is a natural number less than } 6\}$

Practice Exercises

1. $\{5, 10, 15, \ldots\}$
3. $\{x \mid x \text{ is a whole number between 7 and 10, inclusive}\}$

Objective 2

Practice Exercises

5.

Objective 3

Now Try

3a. $-7, 0, 3$
3b. $-7, -\dfrac{8}{9}, 0, 0.4, \dfrac{3}{5}, 2.\overline{15}, 3$
3c. $-\sqrt{3}, \sqrt{7}$
3d. All are real numbers

Practice Exercises

7. (a) 5; (b). $0, 5$; (c). $-4, 0, 5$; (d) $-4, -\dfrac{1}{3}, 0, \dfrac{4}{5}, 5$; (e) $-\sqrt{2}, \sqrt{7}$;

(f) $-4, -\sqrt{2}, -\dfrac{1}{3}, 0, \dfrac{4}{5}, \sqrt{7}, 5$

9. false

Objective 4

Practice Exercises

11. $\dfrac{5}{2}$

Objective 5

Now Try

5a. 11
5b. -10
5c. -10
5d. 9
6. Philadelphia from 2000−2009

Practice Exercises

13. 5

Answers

Objective 6
Now Try
7. >   8. False

Practice Exercises
15. false   17. false

R.2   Operations on Real Numbers

Key Terms
1. product   2. reciprocals   3. difference

4. quotient   5. sum

Objective 1
Now Try
1. −21   2. 3

Practice Exercises
1. −12.4   3. $-\dfrac{16}{33}$

Objective 2
Now Try
3. −22   4. 13

Practice Exercises
5. −3.78

Objective 3
Now Try
5. 12

Practice Exercises
7. 3   9. 9

Objective 4
Now Try
6a. 77   6b. −10

Practice Exercises
11. 3.924

Objective 5
Now Try
7a. −2   7b. −3   7c. 0

7d. undefined

Practice Exercises

13. $-\dfrac{1}{5}$ 15. undefined

R.3   Exponents, Roots, and Order of Operations

Key Terms

1. exponential expression   2. base

3. exponent   4. square root   5. algebraic expression

6. factors

Objective 1
 Now Try

1a. 9^3 1b. $\left(\dfrac{3}{8}\right)^4$ 1c. $(-10)^4$

1d. z^6 2a. 49   2b. $\dfrac{8}{125}$

2c. 256   2d. –32   2e. –625

 Practice Exercises
 1. b^7 3. –8

Objective 2
 Now Try

3a. 7   3b. –30   3c. $\dfrac{4}{9}$

3d. not a real number

 Practice Exercises
 5. 100

Objective 3
 Now Try
 4. 38   5. 34   6. undefined

 Practice Exercises
 7. 20   9. –59

Objective 4
 Now Try
7a. –75   7b. 234

 Practice Exercises

11. $-\dfrac{34}{13}$

Answers

R.4   Properties of Real Numbers

Key Terms

1. coefficient   2. term   3. like terms

4. combining like terms

Objective 1

Now Try

1a. $-12 - 4q$ 1b. $-6r$

Practice Exercises

1. $-3z + 21$ 3. 650

Objective 2

Now Try

2a. $30q$ 2b. $-4 + 5c$

Practice Exercises

5. -2.5

Objective 3

Practice Exercises

7. $\dfrac{1}{5}$ 9. $-\dfrac{1}{7}$

Objective 4

Now Try

3. $-2x - 16$ 4a. $y - 6$ 4b. $24xy$

Practice Exercises

11. $-3x + 17$

Chapter 1 LINEAR EQUATIONS, INEQUALITIES, AND APPLICATIONS

1.1   Linear Equation in One Variable

Key Terms
1. equivalent equations   2. linear equation (first-degree) equation in one variable

3. solution set   4. contradiction   5. conditional equation

6. identity   7. solution

Objective 1
Practice Exercises
1. equation   3. expression

Objective 2
Practice Exercises
5. nonlinear equation

Objective 3
Now Try
2. $\{5\}$

Practice Exercises
7. $\{-5\}$ 9. $\{8\}$

Objective 4
Now Try
3. $\{4\}$

Practice Exercises
11. $\{-8\}$

Objective 5
Now Try
5. $\{-5\}$

Practice Exercises
13. $\{3\}$ 15. $\{60\}$

Objective 6
Now Try
7a. $\{0\}$; conditional   7b. $\varnothing$; contradiction

7c. {all real numbers}; identity

Practice Exercises
17. contradiction; $\varnothing$

1.2 Formulas and Percent

Key Terms
1. formula 　　　　　2. mathematical model 　　　3. percent

Objective 1
Now Try
1. $L = \dfrac{A}{W}$ 　　　　　2. $h = \dfrac{3V}{B}$

Practice Exercises
1. $B = \dfrac{3V}{h}$ 　　　　　3. $y = 2x + 4$

Objective 2
Now Try
5. 60 mph

Practice Exercises
5. 8%

Objective 3
Now Try
6a. 2.6% 　　　　　6b. 3.5%

7. $20.625 million or $20,625,000

Practice Exercises
7. 1100 students 　　　　　9. 15,000 students

Objective 4
Now Try
8a. about 5.4% 　　　　　8b. about 20.1%

Practice Exercises
11. 6.5%

Objective 5
Now Try
9. 0.67 m^2

Practice Exercises
13. 1.06 m^2

1.3 Applications of Linear Equations

Key Terms

1. sum, increased by, more than
2. product, double, of
3. quotient, per, ratio
4. difference, less than, decreased by

Objective 1
Practice Exercises

1. $13 - 7x$
3. $\dfrac{4 + x}{9}$

Objective 2
Now Try

1a. $3x - 100 = 197$
1b. $\dfrac{x}{x + 5} = 23$

Practice Exercises

5. $\dfrac{x}{x - 3} = 17$

Objective 3
Now Try

2a. equation; $\{65\}$
2b. expression; $16x + 31$

Practice Exercises

7. expression; $19z - 59$
9. equation; $\{159\}$

Objective 4
Now Try

3. 6 m, 7 m, 9 m

Practice Exercises

11. Bush: 271 votes; Gore: 266 votes

Objective 5
Now Try

5. $250,000

Practice Exercises

13. $5000
15. $6370.50

Objective 6
Now Try

6. 5%: $1600; 7%: $3500

Practice Exercises

17. 7%: $800; 9%: $300

Objective 7
Now Try

7. 8 oz

Answers

Practice Exercises

19. 8 lb

1.4   Further Applications of Linear Equations

Key Terms

1. supplementary angles   2. complementary angles

3. vertical angles   4. consecutive integers

Objective 1
Now Try

1. 32 quarters; 18 nickels

Practice Exercises

1. 25 large jars; 55 small jars

3. 1500 general admission; 750 reserved

Objective 2
Now Try

2. plane A: 400 mph; plane B: 360 mph   3. $\frac{5}{6}$ hr or 50 min

Practice Exercises

5. plane speed: 265 mph; wind speed: 35 mph

Objective 3
Now Try

4. 20°, 45°, 115°

Practice Exercises

7. 25°, 55°, 100°   9. 30°, 60°, 90°

Objective 4
Now Try

5. 53, 54, 55

Practice Exercises

11. −2, 0, 2, 4

1.5   Linear Inequalities in One Variable

Key Terms
1. interval notation   2. linear inequality in one variable

3. inequality   4. interval   5. equivalent inequalities

Objective 1
Now Try

1a. $(-1, \infty)$;

1b. $(-5, -1]$

Practice Exercises

1. $(3, \infty)$;

3. $(-3, 2]$;

Objective 2
Now Try

2. $(-\infty, -4)$

Practice Exercises

5. $(2, \infty)$;

Objective 3
Now Try

4. $(-\infty, -5]$

5. $(-\infty, -8]$

6. $(7, \infty)$

Practice Exercises

7. $(-2, \infty)$;

9. $(-\infty, 4]$

Answers

Objective 4
Now Try

8. $(-4, -2]$

Practice Exercises

11. $[-5, -3)$;

Objective 5
Now Try
10. 10 feet

Practice Exercises
13. 89   15. 84

1.6   Set Operations and Compound Inequalities

Key Terms
1. union   2. compound inequality   3. intersection

Objective 2
Now Try
1. {30, 50}

Practice Exercises
1. $\{0, 2, 4\}$ 3. $\{1, 3, 5\}$

Objective 3
Now Try

2. [12, 16]

3. $(-\infty, -3)$

Practice Exercises

5. $[-4, 3)$

Objective 4
Now Try
5. {20, 30, 40, 50, 70}

Practice Exercises
7. $\{0, 1, 2, 3, 4, 5\}$ 9. $\{0, 1, 2, 3, 4, 5\}$

Objective 5
  Now Try

6. $(-\infty, 2) \cup [5, \infty)$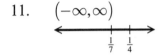

8. $(-\infty, \infty)$

Practice Exercises

11. $(-\infty, \infty)$

1.7   Absolute Value Equations and Inequalities

Key Terms
  1. absolute value equation
  2. absolute value inequality

Objective 2
  Now Try

1. $\left\{-3, \dfrac{7}{5}\right\}$

Practice Exercises

1. $\left\{-\dfrac{13}{2}, \dfrac{7}{2}\right\}$ 3. $\{-2, 14\}$

Objective 3
  Now Try

2. $(-\infty, -2) \cup (1, \infty)$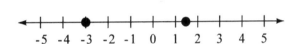

3. $(-3, 2)$

Practice Exercises

5. $(-\infty, -7] \cup [16, \infty)$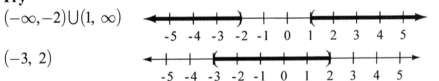

Objective 4
  Now Try
  5. $\{-4, 10\}$

Practice Exercises

7. $\{-2, 3\}$ 9. $\left\{-\dfrac{5}{4}, -\dfrac{1}{4}\right\}$

Objective 5
Now Try
7. $\{-5, -1\}$

Practice Exercises

11. $\left\{-3, \dfrac{11}{3}\right\}$

Objective 6
Now Try

8a. $\varnothing$	8b. $\{-5\}$	9a. $(-\infty, \infty)$
9b. $\varnothing$	9c. $\{4\}$	

Practice Exercises

13. $\{-14\}$ 15. $\varnothing$

Objective 7
Now Try
10. between 16.055 and 17.745 oz, inclusive

Practice Exercises
17. between 16.6465 and 17.1535 oz, inclusive

Chapter 2 LINEAR EQUATIONS, GRAPHS, AND FUNCTIONS

2.1   Linear Equations in Two Variables

Key Terms
1. y-intercept   2. linear equation in two variables
3. coordinate   4. origin   5. x-intercept
6. x-axis   7. quadrant   8. ordered pair
9. y-axis   10. components   11. plot
12. rectangular (Cartesian) coordinate system
13. graph of an equation   14. linear equation in two variables

Objective 3
Now Try

1a. $(0, -2)$ 1b. $(5, 0)$ 1c. $(-5, -4)$

1d. $(10, 2)$ 1e.

x	y
0	-2
5	0
-5	-4
10	2

Practice Exercises

1. (a) $(2, 0)$; (b) $\left(4, -\dfrac{5}{2}\right)$; (c) $\left(-\dfrac{2}{5}, 3\right)$; (d) $\left(0, \ \dfrac{5}{2}\right)$; (e) $\left(\dfrac{2}{5}, \ 2\right)$

3. $(-4, 4), (0, 4), (6, 4), (-12, 4)$

Objective 5
Now Try

2.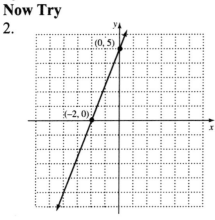

3.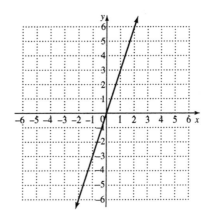

Practice Exercises

5.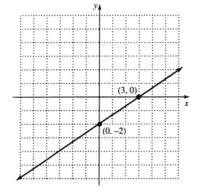

Objective 6
Now Try

4a.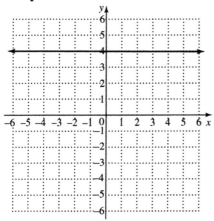

4b.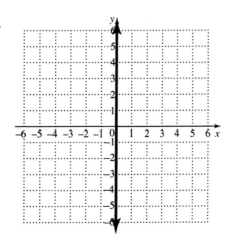

Answers

Practice Exercises

7. $y + 3 = 0$

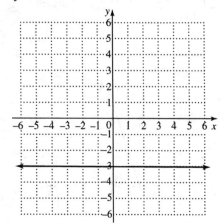

Objective 7
Now Try
5. $(5, -2)$

Practice Exercises
9. $\left(\dfrac{5}{2}, -3\right)$

2.2 The Slope of a Line

Key Terms
1. slope   2. rise   3. run

Objective 1
Now Try
1. $-\dfrac{16}{9}$

Practice Exercises
1. -2 3. $-\dfrac{1}{14}$

Objective 2
Now Try
2. 7   3a. undefined   3b. 0

4. $\dfrac{7}{4}$

Practice Exercises
5. $\dfrac{2}{3}$

Objective 3
Now Try
5.

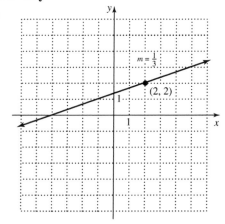

Practice Exercises
7.

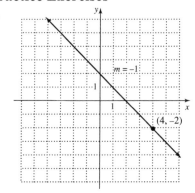

9.

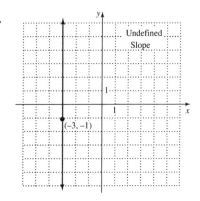

Objective 4
Now Try
6. parallel   7. perpendicular

Practice Exercises
11. parallel

Objective 5
Now Try
9. 113.5 ft/min

Practice Exercises
13. $125,000/yr   15. $950/yr

2.3 Writing Equations of Lines

Key Terms
1. point-slope form   2. standard form   3. slope-intercept form

Objective 1

Now Try

1. $y = \dfrac{7}{9}x + 8$

Practice Exercises

1. $y = \dfrac{3}{2}x - \dfrac{2}{3}$ 3. $y = -\dfrac{6}{5}x + \dfrac{2}{5}$

Objective 2

Now Try

2.

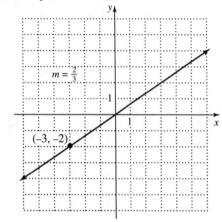

Practice Exercises

5.

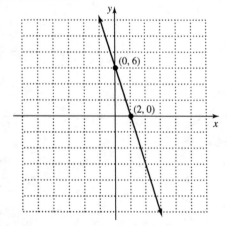

Objective 3

Now Try

3. $y = \dfrac{4}{5}x - \dfrac{44}{5}$

Practice Exercises

7. $3x + 5y = 11$ 9. $2x - 3y = -8$

Objective 4
  Now Try
  4. $y = -\dfrac{3}{4}x + \dfrac{81}{4};\ \ 3x + 4y = 81$

  Practice Exercises
  11. $x + 3y = -1$

Objective 5
  Now Try
  5a. $y = 5$ 5b. $x = -5$

  Practice Exercises
  13. $x = -1$ 15. $y = -5$

Objective 6
  Now Try
  6a. $y = -\dfrac{3}{4}x + \dfrac{7}{2}$ 6b. $y = \dfrac{4}{3}x + 16$

  Practice Exercises
  17. $4x - 3y = -17$

Objective 7
  Now Try
  8a. $y = -28.8x + 686$ 8b. $(1, 657.2)$

  Practice Exercises
  19. (a) $y = 1.25x + 25$; (b) $(0, 25), (5, 31.25), (10, 37.50)$

2.4   Linear Inequalities in Two Variables

Key Terms
  1. boundary line   2. linear inequality in two variables

Objective 1
  Now Try
  1.

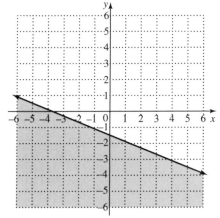

  2.

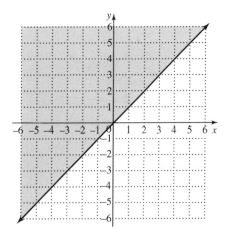

Answers

Practice Exercises

1. $x - y < 5$

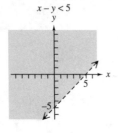

3. $x \le 4y$

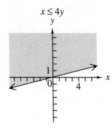

Objective 2
Now Try

4.

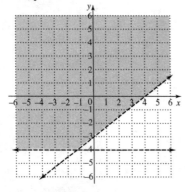

Practice Exercises

5. $4x + y \le 4$
and $x - 2y \le -2$

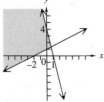

Objective 3
Now Try

5.

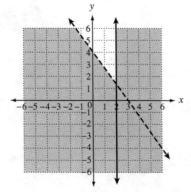

Practice Exercises

7. $4x - 2y \geq -4$
    or $x \geq 1$

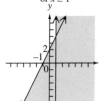

9. $x + y \geq 0$
    or $x - y \geq 0$

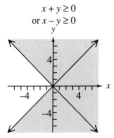

2.5   Introduction to Relations and Functions

Key Terms

1. range

2. relation

3. dependent variable

4. domain

5. function

6. independent variable

Objective 1
Now Try

1. {(1, 3), (2, 4), (3, 6), (5, 7)}

2a. function

2b. not a function

Practice Exercises

1. {(1, 3), (1, 4), (2, −1), (3, 7)}

3. not a function

Objective 2
Now Try

3. domain: {13}; range: {−2, −1, 4}; not a function

4a. domain: $(-\infty, \infty)$; range: $(-\infty, \infty)$

4b. domain: $[-4, \infty)$; range: $(-\infty, \infty)$

Practice Exercises

5. function; domain: {A, B, C, D, E}; range: {V, W, X, Z}

Objective 3
Now Try

5. function

6. function; domain: $(-\infty, \infty)$

Practice Exercises

7. function

9. function; $(-\infty, -6) \cup (-6, \infty)$

Answers

2.6   Function Notation and Linear Functions

Key Terms
1. linear function   2. function notation   3. constant function

Objective 1
Now Try
1. $f(3) = 17$ 3. $g(a-1) = 4a - 11$ 4. $f(-6) = 21$
6. $f(x) = -\dfrac{2}{3}x + \dfrac{7}{3}, f(-3) = \dfrac{13}{3}$

Practice Exercises
1. (a) -13; (b) -7; (c) $-3x - 7$ 3. (a) 9; (b) 9; (c) 9

Objective 2
Now Try
7. domain: $(-\infty, \infty)$; range: $(-\infty, \infty)$

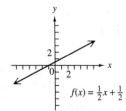

Practice Exercises
5.

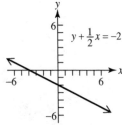

domain: $(-\infty, \infty)$

range: $(-\infty, \infty)$

Chapter 3 SYSTEMS OF LINEAR EQUATIONS

3.1   Systems of Linear Equations in Two Variables

Key Terms
1. independent equations   2. consistent system
3. system of equations   4. solution set of a system
5. dependent equations   6. inconsistent system
7. linear system

Objective 1
  Now Try
  1. yes

  Practice Exercises
  1. not a solution   3. not a solution

Objective 2
  Now Try
  2.

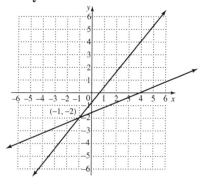

  Practice Exercises
  5.

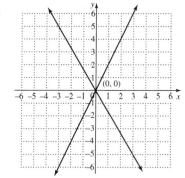

Objective 3
  Now Try
  4. $\{(5,-1)\}$

  Practice Exercises
  7. $\{(-3,-4)\}$ 9. $\{(3, 5)\}$

Objective 4
  Now Try
  7. $\{(-1, 3)\}$

  Practice Exercises
11. $\{(6, 8)\}$

Objective 5
  Now Try
  8. dependent; $\{(x, y) \mid 4x + 3y = 12\}$

9. inconsistent; $\varnothing$ 10. Both are $y = \dfrac{2}{5}x - \dfrac{7}{5}$; infinitely many solutions

Practice Exercises

13. $\varnothing$ 15. $\left\{(x, y) \mid -3x + 2y = 6\right\}$

3.2   Systems of Linear Equations in Three Variables

Key Terms

 1. ordered triple   2. dependent system   3. inconsistent system

Objective 1
Practice Exercises
 1. The planes intersect in one point.

Objective 2
 Now Try
 1. $\{(-3, 1, 2)\}$

Practice Exercises
 3. $\{(-1, 2, 1)\}$ 5. $\{(4, -4, 1)\}$

Objective 3
 Now Try
 2. $\{(2, -5, 3)\}$

Practice Exercises
 7. $\{(4, 2, -1)\}$

Objective 4
 Now Try
 4. $\left\{(x, y, z) \mid x - 5y + 2z = 0\right\}$ 3. $\varnothing$

 Practice Exercises
 9. $\varnothing$; inconsistent system
 11. $\left\{(x, y, z) \mid -x + 5y - 2z = 3\right\}$; dependent equations

3.3   Applications of Systems of Linear Equations

Key Terms
 1. elimination method   2. substitution

Objective 1
 Now Try
 1. length: 17 ft; width: 10 ft

Practice Exercises
1. square: 8 cm; triangle: 13 cm
3. 21 cm

Objective 2
Now Try
2. marigold: $12.29; carnation: $17.60

Practice Exercises
5. large: $6; small: $3

Objective 3
Now Try
3. water: 6 liters; 25% solution: 24 liters

Practice Exercises
7. $6 coffee: 100 lbs; $12 coffee: 50 lb

9. water: 9 oz; 80% solution: 3 oz

Objective 4
Now Try
4. Ashley: 5 mph; Taylor: 3 mph

Practice Exercises
11. distance from school: $1\frac{1}{8}$ mi; jogging time: $\frac{1}{8}$ hr or $7\frac{1}{2}$ min

Objective 5
Now Try
6. 20 lb of $4 candy; 30 lb of $6 candy; 50 lb of $10 candy

Practice Exercises
13. $8 coffee: 15 lb; $10 coffee: 12 lb; $15 coffee: 23 lb

Chapter 4 EXPONENTS, POLYNOMIALS, AND POLYNOMIAL FUNCTIONS

4.1   Integer Exponents and Scientific Notation

Key Terms
1. scientific notation   2. product rule for exponents

3. quotient rule for exponents   4. power rule for exponents

5. base; exponent

Objective 1
Now Try
1. $-24p^6q^6$

Answers

Practice Exercises

1. 7^7 3. $48k^{18}$

Objective 2
  Now Try

2a. 1   2b. 1   2c. −1

2d. 2   3. $\dfrac{1}{(6z)^2}$, $z \neq 0$

  Practice Exercises

5. $2r^7$

Objective 3
  Now Try

5. q^3

  Practice Exercises

7. This expression cannot be simplified further.   9. p^8

Objective 4
  Now Try

6. $\dfrac{-27p^{12}}{q^3}$

  Practice Exercises

11. $-\dfrac{128a^7}{b^{14}}$

Objective 5
  Now Try

8. $\dfrac{4x}{3y^6}$

  Practice Exercises

13. c^{25} 15. $\dfrac{1}{k^4 t^{20}}$

Objective 6
  Now Try

9a. 3.946×10^6 9b. 4.8×10^{-4} 10a. 9,450,000

10b. 0.0000804

  Practice Exercises
17. 0.000042

4.2   Adding and Subtracting Polynomials

Key Terms
1. degree of a term
2. descending powers
3. term
4. trinomial
5. polynomial
6. coefficient (numerical coefficient)
7. monomial
8. degree of a polynomial
9. binomial
10. algebraic expression
11. negative of a polynomial
12. polynomial in x

Objective 1
Now Try
1. $11x^7 - 5x^5 + x^3 - 4$; $11x^7$; 11
2. trinomial; 3

Practice Exercises
1. $y^3 - 7y^2 + 5y + 8$; y^3; 1
3. binomial; 4

Objective 2
Now Try
3a. $14m + 2n$
3b. $-x^2yz + 13xyz^2$
4. $2x^5 - 14x^3 - x$
5. $10y^3 - 12y^2 + 13y$

Practice Exercises
5. $5a^4 - a^3 - 6a^2 + 28a + 1$

4.3   Polynomial Functions, Graphs, and Composition

Key Terms
1. cubing function
2. composite function
3. polynomial function of degree n
4. squaring function
5. identity function
6. composition

Objective 1
Now Try
1. 1

Practice Exercises
1. (a) -7; (b) -17
3. (a) 28; (b) 198

Objective 2
Now Try
2. $10,278.96

Answers

Practice Exercises
5. 236 ft

Objective 3
Now Try
3a. $3x^2 - 6x + 21$ 3b. $9x^2 - 8x + 3$ 4. $7x^2 - x$; 30

  Practice Exercises
  7. (a) $x^2 + 7x - 13$; (b) $3x^2 + x + 3$ 9. 18

Objective 4
  Now Try
  5. -15 7. 76

  Practice Exercises
  11. a. $\frac{1}{5}$; b. 16; c. $\frac{3}{x^2} - \frac{4}{x} + 1$

Objective 5
  Now Try
  8.

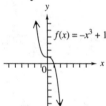

      domain: $(-\infty, \infty)$;
      range: $(-\infty, \infty)$

  Practice Exercises
  13.

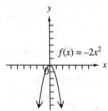

      domain: $(-\infty, \infty)$;
      range: $(-\infty, 0]$

4.4   Multiplying Polynomials

Key Terms
  1. inner product   2. FOIL   3. outer product

Objective 1
  Now Try
  1. $10x^9 y^3$

Practice Exercises

1. $16y^7$ 3. $108r^5 s^7$

Objective 2

Now Try

2a. $-28x^4 - 56x^3 - 42x^2$ 2b. $3x^4 + 15x^3 - 18x^2$

Practice Exercises

5. $6m^2 + 2m - 20$

Objective 3

Now Try

4. $54z^2 - 66z + 20$

Practice Exercises

7. $6x^2 - 5xy - 6y^2$ 9. $x^2 - 2x - 15$

Objective 4

Now Try

5a. $9m^2 - 64y^2$ 5b. $10x^5 - 250x^3$

Practice Exercises

11. $49x^2 - 9y^2$

Objective 5

Now Try

6a. $p^2 - 6p + 9$ 6b. $16p^2 - 56pq + 49q^2$ 7a. $49p^2 - 42p + 9 - 4q^2$

7b. $625a^4 + 500a^3 b + 150a^2 b^2 + 20ab^3 + b^4$

Practice Exercises

13. $25y^2 - 30y + 9$ 15. $4x^2 + y^2 - 4xy + 12x - 6y + 9$

Objective 6

Now Try

8. $42x^3 + 47x^2 + 10x;\ -741$

Practice Exercises

17. 25

Answers

4.5 Dividing Polynomials

Key Terms

  1. dividend   2. quotient   3. divisor

Objective 1

Now Try

  1. $2x^2 + x - 3$

Practice Exercises

  1. $3x^4 + 7x^3 + 5x$ 3. $-\dfrac{r}{3s} - \dfrac{2}{3} + \dfrac{s}{r}$

Objective 2

Now Try

  2. $5r + 7$ 3. $2x^2 + 6x + 3 + \dfrac{2}{x-3}$ 4. $6r^2 - 3r + 7 + \dfrac{-35r+3}{2r^2-5}$

  5. $p^2 - \dfrac{3}{5}p + 4$

Practice Exercises

  5. $9p^3 - 2p + 5$

Objective 3

Now Try

  6. $2x + 5$ $x \neq 4$; 1

Practice Exercises

  7. $4x + 9$, $x \neq 5$ 9. $-\dfrac{6}{5}$

Chapter 5 FACTORING

5.1 Greatest Common Factors and Factoring by Grouping

Key Terms

  1. factored form   2. greatest common factor

  3. factor

Objective 1

Now Try

  1. $6(z - 9)$ 2. $11x(x + 4)$ 3. $(y + 5)(5y - 4)$

  4. $b^2(-7b^2 - 8b + 5)$; $-b^2(7b^2 + 8b - 5)$

Practice Exercises

  1. There is no common factor other than one.   3. $(x + 2)^2$

Objective 2
  Now Try
  5. $(9+r)(a-b)$ 6. $(a-5)(y-3)$ 7. $(p+4q)(9x+1)$
  8. $(a^2-6)(b^2+8)$

  Practice Exercises
  5. $(x-4)(1+2y^2)$

5.2   Factoring Trinomials

Key Terms
  1. factoring   2. prime polynomial

Objective 1
  Now Try
  1a. $(x-7)(x+4)$ 1b. $(p+7)(p+3)$ 2. Prime
  3. $(x+13a)(x-2a)$ 4. $15y(y+2)(y-4)$

  Practice Exercises
  1. $(x+9)(x+2)$ 3. $(rs+7)(rs-3)$

Objective 2
  Now Try
  5. $(3x-1)(6x+5)$

  Practice Exercises
  5. $(2r-3)(10r+1)$

Objective 3
  Now Try
  6. $(4x-5)(2x+1)$ 7. $(5x-3y)(3x+2y)$ 9. $9x(4x-5)(x+3)$
  8. $-(7x-5)(x+2)$

  Practice Exercises
  7. $(3p-5)(2p+3)$ 9. $2ab(a-3b)(a-2b)$

Objective 4
  Now Try
  10. $(3z+7)(4z+19)$ 11. $(4y^2-5)(2y^2+1)$

  Practice Exercises
  11. $(2z-7)(4z-19)$

Answers

5.3 Special Factoring

Key Terms
 1. difference of squares   2. perfect square trinomial

Objective 1
 Now Try
 1a. $(t+13)(t-13)$ 1b. $(5p+8q)(5p-8q)$

 Practice Exercises
 1. $(5a-6)(5a+6)$

 3. $[q-(2r+3)][q+(2r+3)]$ or $(q-2r-3)(q+2r+3)$

Objective 2
 Now Try
 2a. prime   2b. $(m-9+b)(m-9-b)$

 Practice Exercises
 5. $\left(8p^2+3q^2\right)^2$

Objective 3
 Now Try
 3. $(10x-y)(100x^2+10xy+y^2)$

 Practice Exercises
 7. $(2r-3s)(4r^2+6rs+9s^2)$ 9. $(2a-5b)(4a^2+10ab+25b^2)$

Objective 4
 Now Try
 4a. $(r+5)(r^2-5r+25)$ 4b. $(a-5+b)(a^2-10a+25-ab+5b+b^2)$

 Practice Exercises
 11. $(5p+q)(25p^2-5pq+q^2)$

5.4 A General Approach to Factoring

Key Terms
 1. factoring by grouping   2. FOIL

Objective 1
 Now Try
 1a. $18(x+3)$ 1b. $6r^3s(3rs+1)$ 1c. $(y+z)(7x-5)$

 2a. $(11x+5y)(11x-5y)$ 2b. $(3t-4w)(9t^2+12tw+16w^2)$

2c. prime

3a. $(6z-7)^2$

3b. $5(3y-2z)(2y+1z)$

3c. $(4k+1)(k-2)$

4a. $(c+d)(3c^2-d^2)$

4b. $(6b-1+c)(6b-1-c)$

4c. $(5a-b)(25a^2+5ab+b^2+5a+b)$

Practice Exercises

1. $3ab(4ab+a-3b)$

3. $2(4x-y)(16x^2+4xy+y^2)$

5. $(x-3)(x^2+7)$

5.5 Solving Equations by the Zero-Factor Property

Key Terms

1. standard form

2. quadratic equation

3. zero-factor property

4. double solution

Objective 1

Now Try

1. $\left\{-\dfrac{5}{4}, \dfrac{3}{2}\right\}$

2a. $\left\{-4, \dfrac{1}{4}\right\}$

2b. $\left\{\dfrac{3}{7}\right\}$

5. $\left\{-2, \dfrac{5}{3}\right\}$

6. $\{-5, 0, 7\}$

Practice Exercises

1. $\left\{-\dfrac{5}{2}, 4\right\}$

3. $\{3\}$

Objective 2

Now Try

7. width: 10ft

Practice Exercises

5. 18 ft by 30 ft

Objective 3

Now Try

9. $W = \dfrac{S-2LH}{2H+L}$

Practice Exercises

7. $b = \dfrac{c}{d-a}$

9. $w = \dfrac{v}{uv+1}$

Answers

Chapter 6 RATIONAL EXPRESSIONS AND FUNCTIONS

6.1 Rational Expressions and Functions; Multiplying and Dividing

Key Terms

1. rational function   2. rational expression

Objective 1; Objective 2

Now Try

1. $2; \{x \mid x \neq 2\}$

Practice Exercises

1. $\dfrac{2}{3}; \left\{s \mid s \neq \dfrac{2}{3}\right\}$ 3. $1, 2; \{q \mid q \neq 1, 2\}$

Objective 3

Now Try

2. $\dfrac{x-4}{x-7}$ 3. -1

Practice Exercises

5. $\dfrac{y+1}{y-3}$

Objective 4

Now Try

4. $\dfrac{36x}{7}$

Practice Exercises

7. $\dfrac{x-3}{x+2}$ 9. $\dfrac{m-2}{m-4}$

Objective 5

Practice Exercises

11. $\dfrac{z^2-9}{7z+7}$

Objective 6

Now Try

5. $\dfrac{2}{9p}$

Practice Exercises

13. $-\dfrac{2(m-5)}{m+3}$ 15. $\dfrac{4(y-1)}{3(y-3)}$

6.2 Adding and Subtracting Rational Expressions

Key Terms

1. equivalent expressions

2. least common denominator (LCD)

Objective 1
Now Try

1a. $-\dfrac{2}{z^2}$

1b. $\dfrac{1}{c+d}$

1c. $\dfrac{1}{x-1}$

Practice Exercises

1. $\dfrac{4n-7}{m+3}$

3. $\dfrac{1}{k-4}$

Objective 2
Now Try

2a. $33x^2y^4$

2b. $y(y+6)$

2c. $3(x+1)(x-4)^2$

Practice Exercises

5. $3(x+2)(x+3)$

Objective 3
Now Try

5. $\dfrac{x-12}{x-8}$

7. $\dfrac{8x^2+37x+15}{(x+5)(x-5)^2}$

Practice Exercises

7. $\dfrac{-6n^2+27n+12}{(n+4)^2(n-4)}$

9. $\dfrac{y^2-18y-3}{(y+3)(y+1)(y-2)}$

6.3 Complex Fractions

Key Terms

1. complex fraction

2. LCD

Objective 1
Now Try

1a. $\dfrac{5(x+1)}{3(x-1)}$

1b. $\dfrac{2y-7}{7y-2}$

Practice Exercises

1. $\dfrac{5(3a+4)}{2a+5}$

3. $(a+2)^2$

Answers

Objective 2
Now Try

2a. $\dfrac{2y-7}{7y-2}$

2b. $\dfrac{4a^2-8a+1}{3a^2-6a+1}$

Practice Exercises

5. $\dfrac{r^2+3}{5+r^2 t}$

Objective 3
Now Try

3a. $\dfrac{1-r}{r(1+r)}$

3b. $\dfrac{b(b+a^2)}{a(a-b^2)}$

Practice Exercises

7. $\dfrac{7(5k-m)}{4}$ or $\dfrac{35k-7m}{4}$

9. $\dfrac{3(8-x)}{2(15+x)}$

Objective 4
Now Try

4. $\dfrac{1}{xy-1}$

Practice Exercises

11. $\dfrac{r}{s}$

6.4   Equations with Rational Expressions and Graphs

Key Terms
1. vertical asymptote
2. domain of the variable in a rational equation
3. discontinuous
4. horizontal asymptote

Objective 1
Now Try

1a. $\{x \mid x \neq 0\}$

1b. $\{x \mid x \neq -8,\ 7\}$

Practice Exercises
1. (a) $0, -1$; (b) $\{x \mid x \neq 0, -1\}$

3. (a) $-3, 0, 1$; (b) $\{x \mid x \neq -3, 0, 1\}$

Objective 2
Now Try

2. $\{2\}$

3. $\varnothing$

Practice Exercises
5. $\varnothing$

Objective 3
  Now Try
   6.

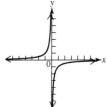

      Vertical asymptote: $x = 0$
      Horizontal asymptote: $y = 0$

Practice Exercises
  7.

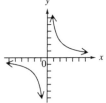

      Vertical asymptote: $x = 0$
      Horizontal asymptote: $y = 0$

  9.

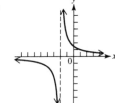

      Vertical asymptote: $x = -2$
      Horizontal asymptote: $y = 0$

6.5   Applications of Rational Expressions

Key Terms
  1. ratio   2. proportion

Objective 1
  Now Try
   1. $\dfrac{11}{3}$

Practice Exercises
  1. $m = \dfrac{75}{8}$ 3. $L = 20$

Objective 2
  Now Try
   2. $R = \dfrac{R_1 R_2}{R_1 + R_2}$

Practice Exercises
  5. $B = \dfrac{2A - hb}{h}$

Objective 3
  Now Try
   5. 4 gallons

Answers

Practice Exercises
7. 8 gal   9. 50,000 people

Objective 4
  Now Try
  6. 2 mph

  Practice Exercises
  11. Pauline: 60 mph; Pete: 40 mph

Objective 5
  Now Try
  8. $\frac{2}{5}$ hour

  Practice Exercises
  13. $2\frac{2}{5}$ hr   15. 2 hr

6.6   Variation

Key Terms
  1. constant of variation   2. varies inversely
  3. varies directly

Objective 1; Objective 2
  Now Try
  1. $25; $P = 25x$ 2. 125 psi   3. 153.86 sq cm

  Practice Exercises
  1. $k = 0.25; y = 0.25x$ 3. 100 newtons

Objective 3
  Now Try
  4. 640 cycles per second 5. $\frac{9}{8}$

  Practice Exercises
  5. 40 lb

Objective 4
  Now Try
  6. 1280 psi

  Practice Exercises
  7. 96   9. 750°

Objective 5
  Now Try
  7. About 63.5 cm^3

Practice Exercises
11. 9 hr

Chapter 7 ROOTS, RADICALS, AND ROOT FUNCTIONS

7.1   Radical Expressions and Graphs

Key Terms
  1. radicand   2. perfect square   3. index (order)

  4. square root   5. radical expression   6. principal square root

  7. radical   8. cube root

Objective 1
Now Try

1a. 4   1b. $\dfrac{4}{3}$

Practice Exercises
1. 5   3. 9

Objective 2
Now Try
2a. −11   2b. −5   2c. not a real number

Practice Exercises
5. 4

Objective 3
Now Try
3a.

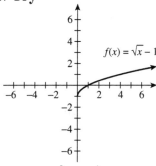

$f(x) = \sqrt{x} - 1$

domain: $[0, \infty)$

range : $[-1, \infty)$

3b.

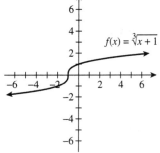

$f(x) = \sqrt[3]{x+1}$

domain: $(-\infty, \infty)$

range: $(-\infty, \infty)$

Answers

Practice Exercises

7.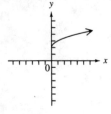

domain: $[0, \infty)$
range: $[2, \infty)$

9.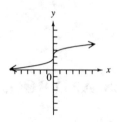

domain: $(-\infty, \infty)$
range: $(-\infty, \infty)$

Objective 4
Now Try
4a. 73

4b. $|n|$

5a. 4

5b. w^{10}

Practice Exercises
11. $-x$

Objective 5
Now Try
7. 23.2 mph

Practice Exercises
13. 8.883

15. 2.2 sec

7.2 Rational Exponents

Key Terms
1. power rule for exponents

2. product rule for exponents

3. quotient rule for exponents

Objective 1
Now Try
1a. 11

1b. -3

1c. not a real number

Practice Exercises
1. -4

3. -15

Objective 2
Now Try

2a. 9

2b. -216

3. $\dfrac{1}{4}$

Practice Exercises
5. 7776

Objective 3
Now Try

4a. $\sqrt[3]{23}$

4b. $\left(\sqrt[3]{2x}\right)^4 - 3\left(\sqrt[5]{x}\right)^2$

4c. $10^2 = 100$

Practice Exercises

7. $4\left(\sqrt[5]{y}\right)^2 + \sqrt[5]{5x}$

9. $a^{1/4}$, or $\sqrt[4]{a}$

Objective 4
Now Try

5a. $5^{5/2}$

5b. $\dfrac{x^2}{y^{1/4}}$

6a. $x^{7/6}$

6b. $x^{1/4}$

Practice Exercises

11. $a^{5/6}$

7.3 Simplifying Radicals, the Distance Formula, and Circles

Key Terms

1. hypotenuse
2. index; radicand
3. legs
4. circle
5. center
6. radius

Objective 1
Now Try

1a. $\sqrt{14}$

1b. $\sqrt{33mn}$

2a. $\sqrt[3]{21}$

2b. $\sqrt[3]{35xy}$

2c. $\sqrt[5]{8w^4}$

2d. cannot be simplified using the product rule

Practice Exercises

1. $\sqrt{42tx}$

3. cannot be simplified using the product rule

Objective 2
Now Try

3a. $\dfrac{6}{7}$

3b. $\dfrac{\sqrt{13}}{9}$

3c. $-\dfrac{7}{5}$

3d. $-\dfrac{a^2}{5}$

3e. $\dfrac{\sqrt[4]{m}}{3}$

Practice Exercises

5. $\dfrac{\sqrt[5]{7x}}{2}$

Answers

Objective 3
Now Try
4a. $2\sqrt{21}$

4b. $9\sqrt{2}$

4c. cannot be simplified

4d. $4\sqrt[3]{4}$

4e. $-2\sqrt[5]{16}$

5a. $10y\sqrt{y}$

5b. $4m^2r^4\sqrt{3mr}$

5c. $-2n^2t\sqrt[3]{4nt^2}$

5d. $-3y^2\sqrt[4]{5x^3y}$

6a. $\sqrt[4]{11^3}$, or $\sqrt[4]{1331}$

6b. $\sqrt[5]{z^4}$

Practice Exercises
7. $\sqrt[3]{x^2}$

9. $5ab^2\sqrt[3]{10a^2b}$

Objective 4
Now Try
7. $\sqrt[6]{63}$

Practice Exercises
11. $\sqrt[8]{28}$

Objective 5
Now Try
8. $12\sqrt{2}$

Practice Exercises
13. 26

15. $6\sqrt{2}$

Objective 6
Now Try
9. $\sqrt{34}$

Practice Exercises
17. 5

Objective 7
Now Try
10. $x^2+y^2=25$

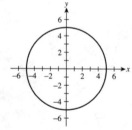

11. $(x+5)^2+(y-4)^2=16$

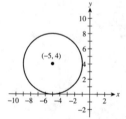

12. $(x-3)^2+(y+1)^2=6$

Practice Exercises

19. $(x-3)^2 + (y+4)^2 = 25$

21. $x^2 + (y-3)^2 = 2$

7.4 Adding and Subtracting Radical Expressions

Key Terms

1. unlike radicals
2. like radicals

Objective 1
Now Try

1a. $-\sqrt{6}$

1b. $13\sqrt{2z}$

1c. cannot be simplified

2a. $-3\sqrt[3]{2}$

2b. $5z\sqrt[4]{2y^2z}$

2c. $6x^2\sqrt[3]{2x} + 6x^3\sqrt{3x}$

3a. $\dfrac{2\sqrt{5}}{3}$

3b. $\dfrac{17}{w^2}$

Practice Exercises

1. $12\sqrt{x}$

3. $\dfrac{8y^2\sqrt[3]{y}}{15}$

7.5 Multiplying and Dividing Radical Expressions

Key Terms

1. conjugate
2. rationalizing the denominator

Objective 1
Now Try

2a. 12

2b. $6 + 3\sqrt{7} - 2\sqrt{2} - \sqrt{14}$

2c. $11 - 6\sqrt{2}$

Practice Exercises

1. $\sqrt{10} - 4\sqrt{5} + 2\sqrt{3} - 4\sqrt{6}$

3. $4 - \sqrt[3]{25}$

Objective 2
Now Try

3. $\dfrac{2\sqrt{15}}{15}$

4. $-\dfrac{3\sqrt{10}}{8}$

5. $\dfrac{\sqrt[3]{10}}{5}$

Practice Exercises

5. $\dfrac{y\sqrt{21b}}{6b}$

Objective 3
Now Try

6. $-2(\sqrt{3}+2)$

Practice Exercises

7. $-4\sqrt{3}+8$

9. $\dfrac{\sqrt{3}+2\sqrt{6}+\sqrt{2}+4}{-7}$

Objective 4
  Now Try

  7. $\dfrac{3+2\sqrt{15}}{4}$

  Practice Exercises

  11. $\dfrac{2-9\sqrt{2}}{3}$

7.6   Solving Equations with Radicals

Key Terms
  1. extraneous solution   2. radical equation

Objective 1
  Now Try

  1. $\{10\}$ 2. $\varnothing$ 3. $\{5\}$

  5. $\{4\}$

  Practice Exercises

  1. $\{11\}$ 3. $\{-9\}$

Objective 2
  Now Try

  6. $\{2\}$

  Practice Exercises

  5. $\{-4\}$

Objective 3
  Now Try

  7. $h=\dfrac{3v}{\pi r^{2}}$

  Practice Exercises

  7. $L=Z^{2}C$ 9. $r=\dfrac{a}{4\pi^{2}N^{2}}$

7.7 Complex Numbers

Key Terms
1. complex number
2. complex conjugate
3. imaginary part
4. real part
5. standard form (of a complex number)
6. pure imaginary number
7. nonreal complex number

Objective 1
Now Try
1. $4i$
2. $-\sqrt{30}$
3. 5

Practice Exercises
1. $-9i\sqrt{2}$
3. $6i$

Objective 2
Practice Exercises
5. imaginary

Objective 3
Now Try
4. $10 - 9i$
5. $24 + 4i$

Practice Exercises
7. $-4 + i$
9. $2 - 3i$

Objective 4
Now Try
6. $-14 + 8i$

Practice Exercises
11. $-8 + 6i$

Objective 5
Now Try
7. $\dfrac{18}{29} + \dfrac{13i}{29}$

Practice Exercises
13. $\dfrac{4}{5} - \dfrac{7}{5}i$
15. $\dfrac{15}{13} + \dfrac{16}{13}i$

Objective 6
Now Try
8. 1

Practice Exercises
17. i

Chapter 8   QUADRATIC EQUATIONS, INEQUALITIES, AND FUNCTIONS

8.1   The Square Root Property and Completing the Square

Key Terms

1. square root property   2. completing the square

3. quadratic equation   4. zero-factor property

Objective 1
Practice Exercises

1. $\left\{-\dfrac{1}{3},\ \dfrac{2}{5}\right\}$ 3. $\{-2, 2\}$

Objective 2
Now Try

2. $\left\{\sqrt{13}, -\sqrt{13}\right\}$, or $\left\{\pm\sqrt{13}\right\}$ 3. About 2.7 seconds

Practice Exercises

5. $\left\{-7\sqrt{2},\ 7\sqrt{2}\right\}$

Objective 3
Now Try

5. $\left\{\dfrac{-5-4\sqrt{2}}{2},\ \dfrac{-5+4\sqrt{2}}{2}\right\}$

Practice Exercises

7. $\{-6, 2\}$ 9. $\left\{\dfrac{-4-4\sqrt{2}}{3},\ \dfrac{-4+4\sqrt{2}}{3}\right\}$

Objective 4
Now Try

7. $\left\{\dfrac{11-\sqrt{89}}{2},\ \dfrac{11+\sqrt{89}}{2}\right\}$

Practice Exercises

11. $\{1, 8\}$

Objective 5
Now Try

9. $\left\{-4i\sqrt{2},\ 4i\sqrt{2}\right\}$

Practice Exercises

13. $\left\{\dfrac{1}{2}-\dfrac{3}{10}i,\ \dfrac{1}{2}+\dfrac{3}{10}i\right\}$ 15. $\left\{1+i\sqrt{2},\ 1-i\sqrt{2}\right\}$

8.2　The Quadratic Formula

Key Terms
1. discriminant　　　　2. quadratic formula

Objective 1
Objective 2
Now Try

1. $\left\{\dfrac{4}{3}, \dfrac{3}{2}\right\}$　　　　2. $\left\{\dfrac{1-\sqrt{7}}{2},\ \dfrac{1+\sqrt{7}}{2}\right\}$　　　3. $\{1-2i,\ 1+2i\}$

Practice Exercises
1. $\{2, 4\}$　　　　3. $\{5-3i, 5+3i\}$

Objective 3
Now Try

4a. 81; two rational solutions; factoring

4b. 0; one rational solution; factoring

4c. -48; two nonreal complex solutions; quadratic formula

Practice Exercises
5. C

8.3　Equations Quadratic in Form

Key Terms
1. standard form　　　　2. quadratic in form

Objective 1
Now Try

1. $\left\{-7, \dfrac{5}{4}\right\}$

Practice Exercises

1. $\left\{-\dfrac{35}{4}, -3\right\}$　　　　3. $\left\{-7, -\dfrac{7}{2}\right\}$

Objective 2
Now Try
2. 2 mph

Practice Exercises
5. 550 mph

Objective 3
Now Try
4a. $\{2\}$　　　　　　　4b. $\{8\}$

　　　413

Answers

Practice Exercises
7. $\{2, 5\}$ 9. $\{9\}$

Objective 4
Now Try
6. $\{-2, -1, 1, 2\}$ 7a. $\{-2, 10\}$ 7b. $\{-1, 27\}$

Practice Exercises
11. $\left\{-27, -3\sqrt{3},\ 3\sqrt{3},\ 27\right\}$

8.4 Formulas and Further Applications

Key Terms
1. quadratic function 2. Pythagorean theorem

Objective 1
Now Try
1. $t = \pm\dfrac{\sqrt{mxF}}{F}$ 2. $q = \dfrac{-k \pm k\sqrt{5}}{2p}$

Practice Exercises
1. $d = \dfrac{k^2 l^2}{F^2}$ 3. $a = \dfrac{-c \pm c\sqrt{2}}{b}$

Objective 2
Now Try
3. south: 72 mi; east 54 mi

Practice Exercises
5. 10 ft

Objective 3
Now Try
4. 2.5 in.

Practice Exercises
7. 2.5 ft 9. 4 ft

Objective 4
Now Try
5. 4.1 sec

Practice Exercises
11. 1.2 sec

8.5   Graphs of Quadratic Functions

Key Terms

1. axis   2. vertex   3. quadratic function

4. parabola

Objective 1; Objective 2

Now Try

1. vertex: (0,–1), axis: $x = 0$;
   domain: $(-\infty, \infty)$; range: $[-1, \infty)$

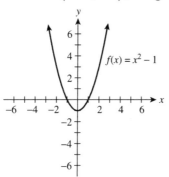

2. Vertex: (–3, 0); axis: $x = -3$
   domain: $(-\infty, \infty)$; range: $[0, \infty)$

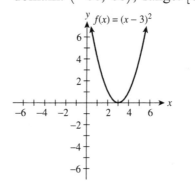

3. vertex: (–2,–1); axis: $x = -2$
   domain: $(-\infty, \infty)$; range: $[-1, \infty)$

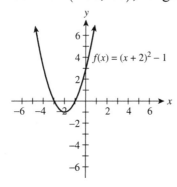

Practice Exercises

1.

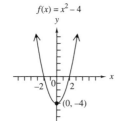

   Vertex: $(0, -4)$
   Axis: $x = 0$
   Domain: $(-\infty, \infty)$
   Range: $[-4, \infty)$

3.

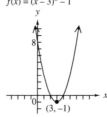

   Vertex: $(3, -1)$
   Axis: $x = 3$
   Domain:
   $(-\infty, \infty)$
   Range: $[-1, \infty)$

Answers

Objective 3
Now Try

4. vertex: (0, 0); axis: $x = 0$
    domain: $(-\infty, \infty)$; range: $(-\infty, 0]$

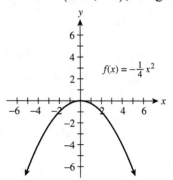

5. vertex: (1, 1); axis: $x = 1$
    domain: $(-\infty, \infty)$; range: $[1, \infty)$

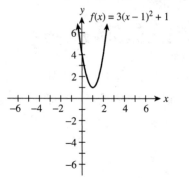

Practice Exercises

5. down; narrower; vertex: (−1, 0); domain: $(-\infty, \infty)$; range: $[0, \infty)$

Objective 4
Now Try

6. $y = 0.01x^2 + 0.77x + 8.36$

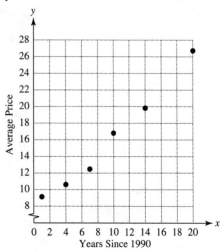

Practice Exercises

7. linear; positive

9.

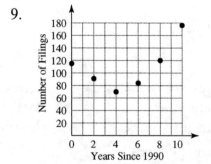

$y = 2.969x^2 - 23.125x + 115$

8.6   More about Parabolas and Their Applications

Key Terms
1. discriminant   2. vertex

Objective 1
Now Try

1. $(3,-5)$ 2. $(1, 1)$ 3. $\left(\dfrac{3}{2}, \dfrac{1}{2}\right)$

Practice Exercises
1. $(1, 3)$ 3. $(-6, 0)$

Objective 2
Now Try
4. vertex: $(-2, 1)$, axis: $x = -2$;
   domain: $(-\infty, \infty)$; range: $[1, \infty)$

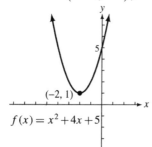

Practice Exercises
5.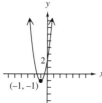

   Vertex: $(-1, -1)$
   Axis: $x = -1$
   Domain: $(-\infty, \infty)$
   Range: $[-1, \infty)$

Objective 3
Now Try
5. 1

Practice Exercises
7. 0   9. 1

Answers

Objective 4
Now Try
6. maximum area: 125,000 sq yd; length: 500 yd; width: 250 yd

Practice Exercises
11. 24 (a square)

Objective 5
Now Try
9. vertex: $(0, 2)$; axis: $y = 2$

domain: $(-\infty, 0]$; range: $(-\infty, \infty)$

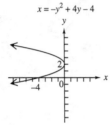

Practice Exercises
13.

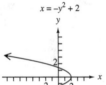

Vertex: $(2, 0)$
Axis: $y = 0$
Domain: $(-\infty, 2]$
Range: $(-\infty, \infty)$

15.

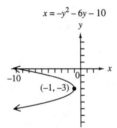

Vertex: $(-1, -3)$
Axis: $y = -3$
Domain:
$(-\infty, -1]$
Range: $(-\infty, \infty)$

8.7   Polynomial and Rational Inequalities

Key Terms
1. rational inequality   2. quadratic inequality

Objective 1
Now Try
1a. $(-\infty, 2) \cup (6, \infty)$ 1b. $(2, 6)$ 2. $(-1, 2)$

4a. $(-\infty, \infty)$ 4b. $\varnothing$

Practice Exercises

1. $[-1, 2]$

3. $\varnothing$

Objective 2
Now Try

5. $\left(-\infty, -\dfrac{3}{2}\right] \cup \left[-\dfrac{1}{3}, \dfrac{1}{2}\right]$

Practice Exercises

5. $(-\infty, -5] \cup [-3, 1]$

Objective 3
Now Try

6. $\left(-\dfrac{3}{2}, -1\right)$

7. $(-\infty, 3) \cup [8, \infty)$

Practice Exercises

7. $(-\infty, 1) \cup [8, \infty)$

9. $[-2, 3)$

Chapter 9   INVERSE, EXPONENTIAL, AND LOGARITHMIC FUNCTIONS

9.1   Inverse Functions

Key Terms

1. one-to-one function

2. inverse of a function

Objective 1
Now Try

1a. Not one-to-one

1b. One-to-one; $G^{-1} = \{(2, 3), (-2, -3), (3, 2), (-3, -2)\}$

1c. Not one-to-one

Practice Exercises

1. One-to-one; $\{(-1, -3), (2, -2), (3, -1), (4, 0)\}$

3. One-to-one; $\{(0, 0), (1, 1), (-1, -1), (2, 2), (-2, -2)\}$

Answers

Objective 2
Now Try
2a. One-to-one   2b. Not one-to-one

Practice Exercises
5. Not one-to-one

Objective 3
Now Try

3a. $f^{-1}(x) = \dfrac{1}{4}x + \dfrac{1}{4}$ 3b. Not one-to-one   3c. $f^{-1}(x) = \sqrt[3]{\dfrac{x+3}{2}}$

Practice Exercises
7. $f^{-1}(x) = \dfrac{x+5}{2}$ 9. Not one-to-one

Objective 4
Now Try
5.

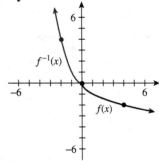

Practice Exercises
11. Not one-to-one

9.2   Exponential Functions

Key Terms
1. inverse   2. exponential equation

Objective 1
Now Try
1a. 8.064   1b. 0.172   1c. 1.246

Practice Exercises
1. 3.737   3. 1.442

Objective 2
Now Try

2.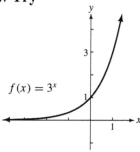

3.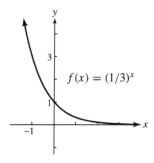

4.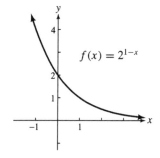

Practice Exercises

5.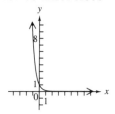

Objective 3
Now Try

5. $\left\{\dfrac{3}{2}\right\}$ 6a. $\{-2\}$ 6b. $\{-5\}$

6c. $\{-4\}$

Practice Exercises

7. $\left\{\dfrac{1}{2}\right\}$ 9. $\{-2\}$

Objective 4
Now Try

7. 256,000   8. about 32,656

Practice Exercises

11. 1 gram

9.3   Logarithmic Functions

Key Terms

1. logarithm   2. logarithmic equation

Objective 1
Objective 2
  Now Try

1a. $\log_8 64 = 2$ 1b. $16^{-1/2} = \dfrac{1}{4}$ 2a. 2.5850

2b. 2.1962   2c. -2.3219 2d. 1.0792

  Practice Exercises

1. $10^{-3} = 0.001$ 3. 2.9299

Objective 3
  Now Try

3a. $\left\{\dfrac{1}{64}\right\}$ 3b. $\{40\}$ 3c. $2\sqrt{3}$

3d. $\left\{\dfrac{1}{6}\right\}$

  Practice Exercises

5. $\{81\}$

Objective 4
  Now Try

4a. 1   4b. 0   4c. 0

4d. 3   4e. 9   4f. 3

  Practice Exercises
7. 0   9. 5

Objective 5
  Now Try

5.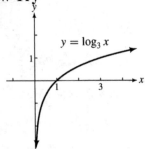

6.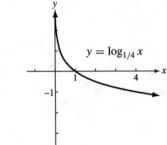

Practice Exercises

11.

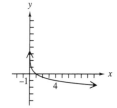

Objective 6
Now Try

7. 16 fish

Practice Exercises

13. 100 mites

9.4   Properties of Logarithms

Key Terms

1. special properties

2. quotient rule for logarithms

3. product rule for logarithms

4. power rule for logarithms

Objective 1
Now Try

1a. $\log_6 5 + \log_6 3$

1b. $\log_5 21$

1c. $1 + \log_4 x$

1d. $4\log_5 x$

Practice Exercises

1. $\log_7 5 + \log_7 m$

3. $\log_4 21$

Objective 2
Now Try

2a. $\log_5 4 - \log_5 9$

2b. $\log_6 \dfrac{x}{3}$

2c. $2 - \log_4 11$

Practice Exercises

5. $\log_6 k - \log_6 3$

Objective 3
Now Try

3a. $3\log_6 4$

3b. $\dfrac{1}{2}\log_b 13$

3c. $\dfrac{3}{4}\log_3 x$

3d. $-7\log_3 x$

Practice Exercises

7. $7\log_m 2$

9. $\sqrt[3]{7}$

Answers

Objective 4
Now Try

4a. $2 + 3\log_6 x$

4b. $\dfrac{1}{2}(\log_b x - \log_b 3)$

4c. $\log_b \dfrac{xy^4}{z}$

4d. $\log_2 \dfrac{x^2(x-1)}{\sqrt[3]{x^2+1}}$

4e. cannot be rewritten

5a. 7.6293

5b. -3.4594

5c. 9.5097

6a. False

6b. False

Practice Exercises
11. 6.1700

9.5   Common and Natural Logarithms

Key Terms
1. natural logarithm

2. common logarithm

Objective 1
Now Try
1. 2.9928

Practice Exercises
1. 1.7576

3. 4.9401

Objective 2
Now Try
2. pH = 7.2076; rich fen

3. 2.5×10^{-4}

4. 85 dB

Practice Exercises
5. 134 dB

Objective 3
Now Try
5. 4.5850

Practice Exercises
7. 4.3347

9. 3.9120

Objective 4
Now Try
6. 11.9 years

Practice Exercises
11. 3240 years

Objective 5
   Now Try
   7. 3.2266

   Practice Exercises
13. 1.1887   15. −2.3219

9.6 Exponential and Logarithmic Equations; Further Applications

Key Terms
   1. continuous compounding   2. compound interest

Objective 1
   Now Try
   1. 2.465   2. 6.770

   Practice Exercises
   1. {1}   3. {−4.324}

Objective 2
   Now Try
   5. $-1+\sqrt[3]{36}$ 7. {1}

   Practice Exercises
   5. {5}

Objective 3
   Now Try
   8. $12,201.90   9. 13.89 years   10a. 5309.18

10b. 34.7 years

   Practice Exercises
   7. $55,200.99   9. 7.7 years

Objective 4
   Now Try
11a. 11.3 mg   11b. 30.0 years

   Practice Exercises
11. 29 years